PROTECT YOUR BUSINESS

Sgt. James Nelson & Ofc. Terry Davis

Small Business Sourcebooks
from Sourcebooks Inc.
Naperville, Illinois

Allen County Public Library
900 Webster Street
PO Box 2270
Fort Wayne, IN 46801-2270

Copyright © 1994 by James Nelson and Terry Davis
Cover design © 1994 by Sourcebooks, Inc.
Porelon ® ink is a registered trademark of Porelon, Inc. This book has been neither authorized nor endorsed by Porelon, Inc.

All rights reserved. No part of this book may be reproduced in any form or by any electronic or mechanical means including information storage and retrieval systems—except in the case of brief quotations embodied in critical articles or reviews—without permission in writing from its publisher, Sourcebooks Trade.

Published by: **Sourcebooks Trade**
A Division of Sourcebooks, Inc.
P.O. Box 372, Naperville, Illinois, 60566
(708) 961-2161
FAX: 708-961-2168

Editorial: John Santucci
Cover Design: Wayne Johnson/Dominique Raccah
Interior Design and Production: Wayne Johnson, Sourcebooks, Inc.

This publication is designed to provide accurate and authoritative information in regard to the subject matter covered. It is sold with the understanding that the publisher is not engaged in rendering legal, accounting, or other professional service. If legal advice or other expert assistance is required, the services of a competent professional person should be sought.
From a Declaration of Principles Jointly Adopted by a Committee of the American Bar Association and a Committee of Publishers and Associations

---

The **Small Business Sourcebooks** series is designed to help you teach yourself the business essentials you need to be successful. All books in the series are available for bulk sales. Feel free to call us at 1-800-798-2475 for information or a catalog. Other books in the series include:

- *How To Market Your Business*
- *Protect Your Business*
- *How to Sharpen Your Competitive Edge*
- *Getting Paid in Full*
- *Your First Business Plan*
- *Smart Hiring for Your Business*
- *The Small Business Start-up Guide*
- *How to Get a Loan or Line of Credit for Your Business*

Library of Congress Cataloging-in-Publication Data

Nelson, James 1948-
    Protect your business / James Nelson and Terry Davis
        p.   cm. -- (Small business sourcebooks)
    Includes index.
    ISBN 0-942061-69-1 : $17.95. -- ISBN 0-942061-66-7 (pbk.) : $8.95
      1. Retail trade --Security measures.   2. Industry -- Security.
I. Davis, Terry, 1956– . II. Title. III. Series
HF5429.27.N45 1994
658.4'73--dc20   93-27225
                                                               CIP

Printed and bound in the United States of America.
Hardcover--10 9 8 7 6 5 4 3 2 1
Paperback--10 9 8 7 6 5 4 3 2 1

Table of Contents

Introduction: Never Underestimate The Power Of The Small Business..........*vii*

Chapter One: The Criminal Justice System *1*

Chapter Two: Embezzlement *13*

Chapter Three: Shoplifting—Anything But Petty *43*

Chapter Four: Armed Robbery *61*

Chapter Five: Commercial Burglary *73*

Chapter Six: Bad Checks *77*

Chapter Seven: Counterfeit Cash *85*

Chapter Eight: Counterfeit Products *91*

Chapter Nine: Charge Accounts *93*

Chapter Ten: Dealing Effectively With Law Enforcement..............................*101*

Chapter Eleven: Firearms In The Business—Requirements And Limitations.....*109*

Chapter Twelve: Robbery And Burglar Alarms *113*

Chapter Thirteen: Technology For
  Prevention/Detection *117*

Chapter Fourteen: Bomb Threats............................*123*

Index..*129*

About The Authors

Sgt. James Nelson has over twenty-two years of law enforcement experience and has worked in patrol, internal affairs, research and development, and investigations. Presently, he supervises an Anti-Burglary Team, Juvenile Investigations, and Crime Prevention Unit. Sgt. Nelson also developed—and currently oversees—a Citizens' Police Academy, which is a highly successful program designed to increase understanding and cooperation between the community and the police department.

In addition to his law enforcement career, Sgt. Nelson is adjunct faculty at Porterville College and instructs courses in the supervision of police personnel and interviewing and interrogation. He holds a B.S. degree in Criminology from California State University, Fresno, and a Master's degree in Criminology from California Lutheran College in Thousand Oaks, California. He is married and resides in Exeter, California.

Terry Davis began his law enforcement career in 1974 and has worked as a police-EMS dispatcher, patrol officer, major crime scene investigator, and as a field training officer. He is presently the coordinator of the Porterville Law Enforcement Training Center's Police Training

About The Authors

Academy. He instructs pre-service courses at Porterville College, covering such areas as crime prevention, criminal investigation, physical evidence, criminal law, and criminal personality profiling. In addition, he teaches ethics, stress management, liability issues, and criminal law at the Complaint-Taker/Dispatchers' Academy at Fresno City College. He has also taught courses in crime scene management and ritual crimes for the State of California and for the U.S. Government.

Mr. Davis also writes a teacher training newsletter, *Techniques For Teaching*, which has subscribers throughout the United States, Europe, and South Africa. He has written a book on physical evidence collection for police officers, investigators, and prosecutors.

Mr. Davis holds a B.S. degree in Criminology and a Master's degree in Education from California State University, Fresno. He is a member of the Golden Key National Honor Society and the International Association for Identification, as well as numerous administration of justice educator associations. He is married, has two children, and resides in Visalia, California.

Together, the authors have over forty years of law enforcement experience. During that time, they have investigated countless crimes committed against businesses and their employees and owners. Cases ranging from simple assault to murder, and shoplifting to elaborate embezzlement schemes, required the authors to learn the operation of the businesses involved. They have logged literally thousands of hours talking with small business owners. As a result, the authors learned a great deal about the problems encountered by businesses today. These experiences, coupled with their investigation and crime prevention expertise, make them uniquely qualified to advise small business owners on how to protect themselves.

Protect Your Business was written as a direct result of having seen too many business owners and employees suffer injuries and business losses, when the injuries might possibly have been avoided and the losses either prevented or minimally reduced. ***Protect Your Business*** is their attempt to help business owners protect themselves.

Introduction

Never Underestimate The Power Of The Small Business!

> *"Even today, with high bankruptcy rates, high unemployment rates and risks in our economy, small businesses are succeeding at an ever-increasing rate. There are almost 20.5 million small businesses in the United States that create two out of three new jobs. They employ about half of the nation's private sector work force and are the major employers of younger, older and female workers. Over half of the inventions we take for granted have come from small companies, such as the zipper, the artificial heart valve, the pacemaker and insulin. Small businesses are helping put a dent in America's foreign trade deficit, contributing about 20 percent of total U.S. exports; in addition, smaller manufacturers are suppliers to larger industrial companies, which are important exporters."*
> —*U.S. Small Business Administration*

More than 99% of all businesses are small businesses, employing some 54.4 million people—*over half of the entire work force*. Small businesses are responsible for 39% of the Gross National Product and 50% of the domestic private sector output. Over half of the sales of products in this nation come through the doors of America's small businesses. And while employment in large businesses has fallen in the past few years,

Introduction

employment in small businesses has posted a steady growth. Between 1990 and 2005, small businesses are expected to be responsible for over 60% of the job growth in the United States.

The small business is the backbone of the American economy. Major corporations may hire more people and pump more money into the local economy. When they fail, however, they fail in a big way. Cities have been known to declare bankruptcy as the direct result of the failure of a single major business or manufacturer. Small businesses struggle through tough times and are the catalyst for rebuilding and renewal. Therefore, anything that can be accomplished to ensure the viability of America's small businesses is worth critical examination.

Many of America's small businesses are in a state of decay and headed toward failure. They are being destroyed by embezzlement from within and burglary and robbery from the outside. Many of these businesses became victims because they failed to make certain practices a part of their everyday activities—practices that would serve to prevent many crimes and minimize the effect of those which were unpreventable.

Protect Your Business is a no-nonsense guide to the prevention of crimes committed against America's small businesses. In today's harsh economic climate, it is an absolute necessity.

Chapter One

The Criminal Justice System

It is not at all surprising that the majority of persons having the opportunity to participate in the processes of the criminal justice system walk away lost in the confusion. To call it somewhat frustrating would be like calling World War II "a minor skirmish."

The police can handle certain things, but not others; the prosecutor won't file charges in a case where the suspect has confessed; the jury recommended the death penalty, but the judge gave the offender life; an offender given a life term walks out of prison after ten years—a free man; a man arrested for robbery/homicide is on parole for a prior robbery/homicide—and this is called a "system?"

It is hoped that as a result of reading this chapter, you will gain a little insight into the roles, responsibilities, and workings of the criminal justice system. Perhaps knowing what to expect will make the frustrations a little easier to handle.

Jurisdiction

Jurisdiction, as it relates to the criminal justice system, refers to both the "territorial range of authority or power" and to the "authority to

The Criminal Justice System

hear and decide cases." For example, a police officer in Tuscon, Arizona is sent to a home within that city to take a report on a robbery. In the process of gathering the necessary information about the crime, it comes to light that the crime did not occur inside the city of Tuscon, but rather, in a small city thirty miles away. This crime occurred outside of the officer's sphere of authority. As a result, the victim of the crime would be directed to make a report to the law enforcement agency representing the city where the crime occurred. In cases where the crime is minor or where the goal is just to document the incident (such as a minor hit and run traffic collision), an agency may elect to take the report as a courtesy, even though it happened in another jurisdiction. This same jurisdictional problem would occur if the crime occurred in the county (an unincorporated area) and the victim wished to file a report with the city police (an incorporated area). Whether termed a"parish" or a "precinct", the agency having jurisdiction over the geographical location where the crime occurred is the agency that should, and in some cases must, take the report and handle the investigation.

Jurisdiction is not always a geographical consideration, for it also pertains to the "authorization to perform." For example, if you are the victim of mail fraud, the local police or sheriff's department may come and take the initial report. They will not, however, investigate the case to fruition. Rather, the case will be turned over to the United States Postal Service and their investigators. They have "jurisdiction" over this type of offense. Transport stolen property across state lines or rob a bank—plan on the Federal Bureau of Investigation becoming involved; this is their area. Where you find a bomb, you will also, at some point, find the Bureau of Alcohol, Tobacco, and Firearms. And, if it's counterfeited cash, it will be the Department of the Treasury.

Often times, this reality is the basis for citizen claims that the police "refused to do anything," when in reality, the police were prohibited from acting, due to geographical jurisdictional boundaries and/or limitations of authority. Obviously, in such cases, referrals should be made directing the victim to contact the appropriate agency. Any law enforcement officer that simply says, "I can't do anything in this case" and leaves it at that, deserves an industrial-strength officer's complaint.

The courts have the same restraints with regard to territorial jurisdiction. A court in one county or parish (or whatever that particular state calls the geographical area) cannot try an individual for a crime that happened in another county (unless the trial has been purposely moved, due to "pre-trial publicity" and in an effort to guarantee a fair trial.) Also included would be the division of responsibility, where certain courts are designated as "criminal courts" and others as "civil courts." One handles matters that are a violation of the written law, while the other handles matters that are societal wrongs, such as one's failure to pay rent, or negligence in the manufacturing of a product that ultimately results in injury or death to the purchaser. There are many different types of courts, ranging from family court (handling divorce cases, child custody matters), juvenile court (which handles only juveniles charged with a criminal or status offense), right up to appeals court (which simply reviews decisions or issues arising from a case in a lower court).

Jurisdiction, of both types, relates to the police, courts, corrections, probation and parole. A more critical examination of jurisdiction will be made as we examine the roles and responsibilities of each of the components of the criminal justice system.

Law Enforcement

The term "peace officer" generally includes a wide variety of classifications of persons. Certainly the most visible is the uniformed beat officer with a city police department. Also included are: sheriff's deputies, state police, highway patrol, parish police, airport police, housing authority police, harbor police, transit police, fire department arson investigators, and maybe even motor vehicles department investigators, district or city attorney's investigators, and a wide variety of state and federal officials.

Incorporated cities generally have their own police department (although some contract these services through the county sheriff). The "jurisdiction" of the local police is generally limited to the confines of the city limits, unless some agreement is worked out otherwise. In the rural, unincorporated areas of the county, it is generally the Sheriff's department who handles law enforcement duties. While they can come into cities having their own police agencies and do have enforcement authority there, they generally limit their involvements to their own areas of responsibility. In addition to the county sheriff, the rural areas may be patrolled by state police or highway patrol. There may be a division of responsibility, in that the highway patrol or state police may have primary jurisdiction for traffic-related offenses (on or off state highways), with the sheriff's department handling all non-traffic related offenses. In another state, the sheriff's department may handle both traffic-related and non-traffic matters.

The Criminal Justice System

The primary role of the police is to protect lives and property through prevention and deterrence of crime and to detect crime when it does occur and apprehend those persons responsible. This is a far cry from the totality of their involvements. The majority of police calls for service are of the non-emergency type, and generally involve some sort of a civil disturbance. Cops handle everything from full blown family fights to "barking dog" complaints. And when the first rains of the season bring up all that oil soaked into the roadway, cops work accidents—lots of accidents. The reality is that the calls for service (of all types) keep patrol officers so busy, that there is seldom time for them to initiate any investigations on their own. Some officers describe their role as "putting out fires," i.e., rushing from place to place, putting out small blazes, and having no time to look for the arsonist. Sometimes it is overwhelming and somewhat similar to bailing out a sinking ship with a teacup (while making no attempt to plug the hole!).

Through their crime prevention role, law enforcement agencies have branched out into a variety of educational areas. No longer is crime prevention limited to teaching the public how not to become a victim. It now includes the community relations aspect, as well, with the theory being that the better the relationship between the agency and the citizenry, the less likely the citizenry is to offend and the more likely they are going to feel compelled to come forward with information in the event a crime is observed by them. Law enforcement agencies are now involved in the education of elementary school children, such as the Drug Abuse Resistance Education program (D.A.R.E.), which originated in the Los Angeles Police Department. Police Athletic Leagues and other such organizations which seek to link law enforcement and youth are productive ways to have an effect upon the escalating crime problem. While the uniformed patrol officer driving the marked patrol car is most assuredly the most visible member of the department, there are many whose involvements are behind the scenes, yet every bit as important.

The communications division of a law enforcement agency is composed of complaint takers (telephone operators) and radio dispatchers. Police complaint takers are the first contact that most victims and witnesses have with law enforcement. Often the callers are highly excited or even injured, and it is the job of the complaint taker to elicit as much information as possible to pass along to the dispatcher (who in turn will give the pertinent data to the uniformed patrol officer). They constantly re-prioritize calls, giving greater concern to crimes in progress than old or "cold" crimes, and higher priority to crimes against persons over crimes against property. It is an extremely difficult job. In small agencies, a single person may be the complaint taker and the dispatcher. Additionally, that same person may also be responsible for typing, filing, teletype requests, computer data entry, preparing paperwork for court processes, helping persons at the counter, and report writing. Whether working for a large agency where their duties may be highly specialized or for a small agency where they are expected to know and do most anything, communications personnel are most assuredly, the unsung heroes of the law enforcement profession (much like football linemen and registered nurses). Their role in every police action is crucial, yet when the praise is given, it usually goes to the field patrol officer.

Uniformed patrolmen (often referred to as "beat cops") seek to both prevent crime (by their presence) and detect crime. All too often, however, their role is reactionary, where they spend the majority of their shift running from one call to the next, writing reports of the crime that has already occurred. There is little time to try and prevent the crimes, or to catch them in the act (although such does still occur). This is the basis (arriving after the fact; documentational role) for much of the frustration felt by street cops. While many police agencies strive to be more pro-active (i.e., go after the offenders before they victimize scores of persons), they are often restrained by funding, staffing, and an agency head that cannot politically afford to allow a reduction in what are truly non-essential police services (like barking dogs, skateboard ramps left in the roadway, neighbor's trees dropping walnuts on the caller's yard, and so on).

Generally, their role in a criminal action is that of the preliminary investigator. They respond to the scene and stabilize whatever is occurring there. They will coordinate any attempt to locate and apprehend the offender(s). It is their responsibility to document what has taken place and who was involved. They will seize the evidence and sometime during that shift write a report of their investigation. (Note that few law enforcement television programs show officers sitting for an hour or more and writing reports of what he or she found, did, didn't do, learned, and proved impossible—it would make for pretty boring footage!). In most cases, the field officer does not conduct any further investigation, but instead, will forward the report to the detective division. Small agencies often have the uniformed "beat cop" handle the case from beginning to end, simply because there is nobody else to call! Thankfully, not all cases require investigation beyond that which the field officer has accomplished.

The Criminal Justice System

The role of the uniformed officer cannot be overstated. They have more to do with the success of a criminal investigation than anyone who comes later in the investigation. It is the uniformed officer who has the best opportunity to collar the crook and obtain the evidence necessary to make the charges stick. If the uniformed officer fails to secure the crime scene to prevent contamination, there's little the crime lab can do with it, despite all of their bells, whistles, and lasers. If the uniformed cop doesn't identify the witnesses and keep them at the scene, follow-up detectives are going to need more than a magnifying class to find that proverbial clue. It isn't by accident that the patrol division is known as the "backbone of the police department."

The detective division (in some agencies, it is called "investigations") is responsible for taking the report written by the field officer and conducting a "follow-up." (It is important to note that it may be several days before the case is assigned to an investigator.) This "follow-up" generally includes an additional interview with the victim(s) and witness(es) to verify the information in the report and to identify any new information that has come to light since the officer left. If there is a suspect in custody, the detective may conduct an interview, where the suspect is read his rights and, if he opts to waive them, a statement is obtained. The detective may become involved in showing photographic line-ups (usually a series of six pictures of which one is the suspect). The detective may request and arrange for forensic examinations of the evidence. In addition, the detective will work closely with the prosecutor to build a case strong enough for a conviction. It is not uncommon for a detective to have thirty to forty cases pending at any given time—there is no luxury of handling one case to its end and then starting on the next one. In addition to following up on existing reports, effective detectives will initiate their own case investigations based upon their own observations or upon information received from developed sources. Good detectives get calls about crimes that have occurred—lots of them.

Detective divisions are generally divided into areas of specialty. In most agencies, you will find divisions that work on crimes against persons, crimes against property, and miscellaneous crimes. These divisions may be further broken down into homicide, robbery, sexual assault, burglary, narcotics, forgery/fraud, auto theft, civil, etc., depending upon need. Sheriff's departments that work rural areas may have a cattle theft division (yes, it still is big business).

On the state level, law enforcement officers generally work within their designated area of responsibility. They may initiate an investigation into any matter or situation that falls under their purview. For example, the state bureau of alcoholic beverage enforcement performs follow-up investigations on complaints/violations of law submitted to them by local or county law enforcement. But, they certainly have the right to enter any licensed premises to see for themselves if state law is being followed. This is true of most state agencies.

Prosecutors and Defenders

The prosecutor represents the people in a criminal action and is, in essence, the attorney for the victim of the crime. The prosecutor may be an employee of the city (city attorney), county (deputy district attorney), state (assistant state's attorney) or a federal prosecutor. Although the police have already investigated the crime, the prosecutor may meet with detectives and officers who are involved in the case and request that additional investigation be conducted. The prosecutor knows where the loopholes are and what problems may be encountered in court. Also, the prosecutor may have a staff of investigators that he will use to conduct follow-up investigations. There is often a wide gap between what you suspect and what you can prove in a court of law. The county district attorney is generally identified as the "chief law enforcement officer of the county." In actuality, the district attorney is an arm of the law enforcement component, not the judiciary.

The defense counsel may be a private attorney hired by the accused, or appointed by the government and paid for by the government. Many county jurisdictions have "public defenders" that are county employees and whose job is to provide counsel and a defense for those charged with crimes. It is possible that the accused may choose to represent himself (called *pro per* and *pro se*), which is generally an ignorant move and has more to do with the offender's ego and need for control than any actual belief in his or her abilities. While it is easy to despise defense attorneys for representing the guilty people arrested by police officers, one must remember that their role is primarily to ensure that the accused receives all due process rights afforded him by the United States Constitution. Everyone is entitled to their day in court and if the cops did their job, justice normally prevails. When it doesn't, it is usually due to weak evidence (often in the form of poor testimony from witnesses—including cops), or because the jury erroneously placed too much emphasis on factors having nothing to do with the question of guilt or innocence. How can defense attorneys represent someone they know is guilty? It's easy—they get paid to do it. The guy who pumps out the septic tank doesn't necessarily like that either. The truth is, many (if not most) crooks lie to their attorneys just like they lie to the police, probation officers, and yes, even mom.

The Criminal Justice System

Defense attorneys may attempt to contact you to obtain a statement or simply to confirm what you have already told law enforcement investigators. Despite their acting ability, recognize that they are seldom out for justice and more or less are out to protect their win/loss record with little concern for the victim in the case. If defense attorneys are vultures, then private investigators hired by attorneys eat what the vulture won't touch.

Private investigators are often hired by defense attorneys to try and interview victims and witnesses (including cops) prior to the trial. Although they will slickly tell you they are only interested in truth and justice (they leave out "the American Way"), the real truth is that they are looking for something to use against you later in court to trip you up and damage your credibility. They are hoping you will make a statement that will show you are less than certain about an identification of their client as the crook, or were perhaps coerced by the police into making an identification. They will try to get you to make an inconsistent statement that they can throw in your face later on in court.

The truth is that you have no obligation to talk to any private investigator nor to the defense counsel about the case and it is NEVER in your best interests to do so. Politely say that you will be more than happy to answer any questions they have *in court*. (Of course, they will try to suggest that you can avoid being humiliated in court if you will just answer a few questions, now. If they knew anything with which to humiliate you they wouldn't be trying to get you to talk to them!). If you have a defense attorney and/or a private attorney that is persistent and will not take "no" for an answer, simply hang up the phone or close the door—whichever is applicable. Then, notify the police and the prosecutor's office that you are being harassed. This type of harassment is unethical at best, and may even violate the law. Judges take a very dim view of anyone who attempts to circumvent the system of justice by intimidating or harassing witnesses (especially if it is the suspect or someone acting on behalf of the suspect, and especially if threats are made).

In one child-molestation case, the private investigator was so slick, that he managed to get invited into the victim's home for a beer, and later, dinner. He then used everything told to him to humiliate the people when they took the witness stand. Worse yet, the pervert walked and a five-year-old little girl didn't understand why. Talking with a private investigator hired by the defense counsel is never in your best interest. Tell them that you already gave a statement and that they can get it from the police officer's report.

The Criminal Justice System

The Criminal Courts

While there are many types of courts (traffic courts, trial courts, justice courts, municipal courts, magistrate's courts, civil courts, family courts, probate courts, appellate courts, etc.), we will only be concerned with the trial courts that a business owner may be summoned to as a witness or as a victim.

As a general rule, most local criminal courts are arranged in "tiers." This means that there are lower-level and upper-level courts. The minor offenses are most often heard and disposed of in the lower-level courts. More serious offenses, however, may start in lower courts for preliminary hearings, evidence-suppression hearings, etc., then move up to the higher courts for the actual trial. An individual tried and convicted in a local criminal court may appeal his case (on some issue of due process or question of law), which will place the case onto a second tier of courts—this time at the state level. It is conceivable (although its actual frequency is quite limited) that the issue of the appeal could reach into the third tier of courts, at the federal level. The issue may ultimately be decided by the Supreme Court of the United States (after which there is no other point of appeal, hence its designation as "the court of last resort").

The appeals process is not the only route by which State and Federal courts may hear or become otherwise involved in a case. State and federal courts can exercise "original jurisdiction" rights and request that the local case be sent up to their level. As well, some offenses automatically are handled at the state or federal level. The state prosecutor's office may try a local case because of a conflict of interest at the local level (often situations where the prosecutor has represented the accused prior, when in private practice or when it's one of their attorney's kids). And, a person convicted of cutting wood without a permit in a National Park would appear before a federal judge in federal court (much to their surprise!), simply because they have original jurisdiction in that part of the country.

It is highly recommended that you consider enrolling in a course on the criminal justice system of your state at your local community college. It may assist in making a highly confusing, often very frustrating "system" much more understandable and useable.

The Civil Courts

Civil courts are generally arranged in tiers as well. The civil courts are generally divided by loss amounts (called an "amount in controversy"). For example, a lower court may only accept civil actions where the initiator (generally called the plaintiff) is suing for an amount below $5,000. Any lawsuit above this amount would have to be filed in a higher court. Again, the numbers chosen here are merely for the sake of example. It is recommended that business owners query their local chambers of commerce to obtain further information about filing suits in local courts.

The Marshal's Office

In some states, there are county-funded marshal's offices that provide for security in the courtrooms and also serve civil papers, enforce evictions, and seize property. They are mentioned here because you may be dealing with the Marshal's Office to seek enforcement of court awards after a civil trial. In some jurisdictions they may be known as a "constable." There are also federal marshals that are responsible for the transportation of federal prisoners and serve a variety of functions at the federal level.

Corrections

When people speak of corrections, they are generally including local, county, state, and federal jails, and probation and parole (county, state, and federal).

Local and county jails generally house prisoners for shorter durations than state or federal facilities (the crimes for which they are incarcerated are less serious). Local jails are usually utilized to house prisoners during processing or while awaiting transfer to the local county jail. Some cities even have "sober up" facilities known as detoxification centers. (A drunk tank smells the same by any other name.) County jails house prisoners while they are awaiting trial and those who have been sentenced to serve punitive time at that level. Federal correctional institutions house prisoners who are charged with and serving time for crimes that violated federal laws.

Bail is a procedure that is limited to local and county jails—one does not "bail" out of state prison lest they tunnel out. Bail is nothing more than money or property being turned over to the city or county as a good-faith exchange for freedom until the trial date. Prisoners may post the bail themselves, in cash. The prisoner, upon appearing in court, would get all of the money back. Failure to appear, however, would result in the forfeiture of the bond money to the authority. Prisoners may instead utilize the bail-bond system, where the total amount of bail is turned over to the proper authority by the bondsman, and he or she, in exchange, keeps a portion of the fee paid, usually around 10 to 15 percent, as their profit for taking the risk that you won't show up (in which case, the bail bondsmen would lose all of their money; this is when "bounty hunters" get phone calls!).

In some cases, offenders arrested and jailed for criminal offenses are released on their "own recognizance" or "OR'd." At the first appearance before the magistrate (judge), they are deemed trustworthy and dependable and are released on their promise to appear for future hearings. There are many victims who wouldn't have been victims if judges hadn't released the offender on his "good name and character." However, judges are aware that the jails are grossly overcrowded (often because of their own orders), and will most likely be releasing all but the most violent offenders, anyway. It's a risk they feel they must take. Unfortunately, the public safety is sometimes jeopardized. The primary considerations of granting bail or an OR release are the severity of the crime, the likelihood that such criminal acts will continue, the safety of the public, and the likelihood that the accused will, in fact, show up in court, rather than in Rio.

Probation is the monitoring of offenders who, generally because their crimes were minor in nature, were released after serving local jail time or in lieu of jail time. A popular method of "monitoring" today is through the use of ankle bands that verify, through a phone call generated by the agency's computer at random times, that the offender is where he is supposed to be. The caseloads of probation officers have grown dramatically, and any thought of close monitoring and counseling is becoming a thing of the past. Rather, offenders refer to the computer as being their probation officer.

Basically, if you don't get in trouble again while on probation, you have little contact with the probation authorities. This is especially sad, because probation (and parole) agents have the ability to work with offenders' charges and help them turn their lives around—there just isn't enough staff to allow that kind of one-on-one counseling any longer. It is not uncommon for a probation officer to have a hundred or more cases on his caseload. Some offenders even commit additional crimes before the probation officer gets a chance to meet with them.

The Criminal Justice System

Parole is a state or federal entity, as one cannot be "paroled" from serving local time (it must be time served in a state or federal prison, generally for crimes that are more serious in nature). Parole, too, is a wonderful idea that simply hasn't worked. Agents are so overburdened with massive caseloads that close monitoring is impossible. The sad truth is that the majority of offenders released on parole re-offend and are sent back to prison. Parolees are generally required to report in to the parole agent at specified times, may not leave the area without written permission, and may have to take routine tests of their urine or blood for the presence of drugs. The reality of the situation is that, as soon as they are released, they head back to the same "digs" and hang with the same guys and do the same things that got them to prison in the first place. This is the reason why we read so often about a murderer on parole committing murders. In one state, a murderer was allowed to go home for Thanksgiving dinner. He committed two more murders between the turkey and the pumpkin pie. Where was the public safety factor in that decision?

In both parole and probation cases, the court may place specific requirements on the individual, such as not to drink alcohol, use drugs, associate with certain people, go to certain places, or be out after specified hours. Violation can result in revocation of parole and the parolee being sent back to prison. Most offenders who are substance abusers (a problem that in most cases got them to prison in the first place) usually violate the terms of their parole after their first scheduled meeting with the parole agent. By absconding (failing to show), they prevent their urine test from coming back from the lab as "dirty," but also they manage to pick up a parole revocation warrant in one fell swoop.

Chapter Two

Embezzlement

The crime of embezzlement is the fraudulent appropriation of property by someone lawfully entrusted with its possession. Basically, it is the taking of the money or property of another, where the offender acquired possession legally via some form of employment or position of trust.

Embezzlement generally comes in one three forms: (1) the direct theft of cash money; (2) the theft of inventory; or (3) the manipulation of accounts. Typically, embezzlement involves the theft of moderate amounts of money over a long period of time, with a high cumulative loss (in one case, an employee who was caught admitted to having embezzled small amounts of money from the company each workday for over eighteen years!)

Just how widespread embezzlement *is* differs with the study. Rest assured that any study based upon crimes reported to police will be erroneous. Most embezzlement cases are never discovered—or at least, never linked to a particular employee. They are difficult cases to work. Paper trails are often time-consuming and, if the victim business has sloppy, insufficient, or non-existent records, the likelihood of the case ever being solved is low. When records are available, it is generally an

Embezzlement

accountant or auditor that pours over the documentation, not a police officer (who is generally ill-equipped for such investigations). Cash is very easy to steal and very difficult to trace. Of those cases that are linked to an employee, few are ever prosecuted. Many business owners are not anxious to report that they allowed an embezzlement to occur—it is negative advertising to potential or long-standing customers. Certainly, any bank or other financial institution would not want the local newspaper to report that an embezzlement had occurred. As well, there is the "good ol' boy syndrome," where employers are hesitant to prosecute long-term employees or even short-term employees who can provide a logical explanation for their acts, such as an illness in the family, trying to make ends meet, etc.

There is a generalized tolerance of those who embezzle. Most business owners are quick to fire, but slow to prosecute. Jurors deciding embezzlement cases tend to be tolerant, as well. Maybe it is because the offenders don't look like crooks—they look like you and me. And, maybe they had a reason to steal—a necessity. Embezzlement is sometimes viewed as a "Robin Hood" story, with a poor worker taking a moderate amount of cash from a tyrannical, filthy rich, corporation. The tolerance extends to judges, as well. In one study, it was found that 91% of armed robbers went to prison regardless of the small amount of cash taken. Embezzlers of large amounts of money and/or property, however, went to prison only 17% of the time.

The Federal Bureau of Investigation reported that they have documented a 40% increase in embezzlement since 1985. This statistic is misleading, at best. While it could mean that employees are more willing to steal from their employers, it could also be indicative of an increased willingness on the part of employers to report employees who steal from them. As well, it could mean nothing more than law enforcement has taken a more aggressive role in prosecuting such cases. One thing that is well-known throughout the industry of asset protection is that far more money and inventory goes out the door in the hands, pockets, and property of employees than is taken by shoplifters—about five times more!

It is estimated that about one-quarter of all workers are totally honest at all times. Another quarter are totally dishonest whenever possible. The remaining 50% are as honest as controls and personal motivation dictate—meaning they'll go either way, depending upon the situation and the need. The generally accepted figure is that approximately one-third of all employees are ripping off their employers to a tune of some $5-10 billion dollars a year.

Embezzlement

Why, in this day of joblessness, would employees steal from their employer at a risk of losing their job? The bottom line is that no thief ever believes he will be caught. Studies have clearly shown that the most predictive factor of theft involvement was the employee's perception of getting caught. The greater the risks, the lesser the chances of theft. If they truly thought they would be caught, they would not commit the theft.

Some employees feel that they are above reproach. Because of their time with the company, their status, or to whom they're related, they feel they would never be accused of theft. Most feel they are too clever to get caught. And, while their thefts often do show planning and some degree of ingenuity, most embezzlement methods are the same.

There is also a rationalization that occurs that allows the embezzlers to convince themselves that they are owed or due the money or product taken. The first thing that comes to mind is uncompensated overtime—the many times they worked overtime and were never paid anything additional. They think about the times that they had to pay, with their own money, for something that should have been paid for by the company. Perhaps it was postage-due mail that the employee paid for out of his or her own pocket and never sought reimbursement for. Employees think about the time they had to order lunch in because the business was too busy to allow them to leave, or the time they paid for a motel room and dinner when traveling on company business. They convince themselves that the company can afford the loss, working under an assumption that business owners rake in ridiculous profits and exploit both the workers and the patrons alike. They convince themselves that the taking of money or property was justified and that it really was owed to them.

Employees may *borrow* money with the intention of paying it back later. The money may be used for almost anything from speculating in the market to paying for narcotics. The key is that the money is taken with an intention of repaying it later. Of course, the money is never repaid and, if he or she is never caught, the employee has now found a valuable source of money.

There are as many reasons for embezzlement as there are stars in the sky. Some are a little more convincing than others. When one suffers financial hardships and cannot see any way to increase the money coming in, they may resort to theft. Illness can devastate a family budget in the blink of an eye. These are known as "non-sharable" problems. Some employees live beyond their means. If they're driving new cars, wearing new clothing, sporting expensive jewelry, living in expensive "digs" and travel a great deal, and you know they only make a thousand dollars a month, a little suspicion is in order.

A common precursor to embezzlement is substance abuse. When an employee becomes addicted to anything, there is a paradigm shift—everything resets to zero. Now, your most trusted employee (or for that matter, partner) cannot be trusted to work alone or to have a key to the door. The individual whom you know would never steal from you will steal you blind and drive you into bankruptcy. Most employers are shocked that any employee would steal from them and often deny even the possibility that certain employees, because of their trusted status, would ever steal from them. "There must be some other explanation," is a common response when employers are confronted with the truth. Remember, if an employee becomes addicted to alcohol, gambling, or illicit drugs, that employee's "morality and ethics tapes" have been erased and replaced with ones that simply say "take whatever you need, the end justifies the means." If young children addicted to cocaine can steal their parents' prized possessions and sell them for a quarter of their actual value (not to mention sentimental value), then certainly an employee in the same set of circumstances can be expected to rip off their employer.

Hiring Trustworthy Employees

Most business owners agree that finding and keeping good employees is the biggest challenge they face. So it is little wonder that business owners sometimes settle for employees that are less than ideal. Before you settle for someone, keep in mind that businesses lose more to employee theft than robbery, burglary, and shoplifting combined.

The problem of employee theft is certainly more complicated than just picking the right person for the job. Management and supervision styles contribute as much as anything else to employee theft. Employees that feel like a part of the business and that are treated fairly are less likely to steal. Job satisfaction is highly correlated with involvement in theft from the employer. Employers with strong inventory systems and who are on top of the business lose less to internal thefts than those with a more casual approach.

It is extremely important that you send clear and consistent messages to your employees. Have a set of rules for employees to follow and then enforce them. If you allow employees to use the telephone, stationery, or other company assets or equipment for personal use, make sure you have very clear guidelines for them to follow.

By becoming an investigator for about two hours, small business owners can prevent most of their internal thefts problems. A thorough background investigation into an applicant's work history, credit, and ties to the community can go a long way in preventing theft.

The Background Investigation

Check with your local credit bureau to find out what the state laws are for obtaining credit histories on job applicants. The law will vary from state to state. In most states, you will be required to obtain a waiver from the applicant before you can view their credit history. The credit bureau can generally supply you with the proper forms that the applicant will have to sign. If you feel uncomfortable about asking an applicant for a waiver, just tell them that your insurance company requires it.

Once you receive the information, review it closely. If the applicant owes a lot of money and is behind on most of her bills, this is a clear signal of warning. The temptation to embezzle will certainly be greater for someone who is in debt. It also demonstrates a lack of responsibility on the applicant's part, which may say something about the type of employee you'll be getting.

If the applicant's credit is marginal, you may want to investigate further. The applicant may explain that the bad credit is the result of illness, divorce, or some other unusual occurrence. These are also popular excuses, so ask for supporting data. They should be able to show you doctor bills or legal papers that will support their claim.

If the employee just got out of high school or college, it is very likely that no credit history exists. If this is the case, you must take a very close look at the applicant. Does the applicant's appearance (clothing, jewelry, vehicle, etc.) point to one who is living well beyond her means? Does the applicant's vehicle have current registration? Where does the applicant live? Will the salary you are offering support the applicant's lifestyle? Many of these questions can be asked in a friendly manner during the first interview. By asking questions about hobbies, long term goals, and recreational activities you can learn a great deal about another's lifestyle.

Embezzlement

The Job Application

Every prospective employee should complete a detailed application. An application that is complete and neat can tell you a lot about what kind of job you can expect from this person. If someone lies on an application, they may be likely to steal from you.

The job application should minimally include:

1. Name, address, date of birth, phone number, driver's license and social security numbers.
2. Spouse's name and work address.
3. Parent's name, address, and phone.
4. Names, addresses, and phone numbers of three persons they have known for at least five (5) years.
5. A listing of all schools attended, including dates and degrees or diplomas received.

This information can be verified by calling the phone numbers provided. You should tell those with whom you speak that "Mr. X" or "Ms. X" has applied for a job with your business and that you are merely checking the references. If you feel uncomfortable about the calls, you can always say that your insurance company requires the verification of all applications. It is unlikely that anyone listed as a reference will have anything bad to say about the applicant. Have you ever used someone as a reference that you knew would say something bad about you?

Besides verifying the information, you want the applicant to know you have made these calls. Certainly the applicant will hear from relatives that you have called. By making these calls, you are sending a clear message that you are thorough and look for the smallest details. This creates a mind-set in the potential employee that you will not be an easy mark. If this employee should steal from you later, the information you have obtained in the background check will be valuable to police, especially if they have to search for the individual.

Unless your business requires some special education or training, you may not need to call the schools listed on the application. However, you may want to contact some of the applicant's teachers. Teachers can often provide a lot of insight about a past student's work habits and citizenship. You should also be careful of anyone that appears too qualified. The applicant has either overstated his qualifications or, for some reason, has been unable to acquire a job for which she is qualified.

With regard to experience, the job application should also ask the applicant to:

6. List all employment for the last five years. Include dates, salary, and reason for leaving.

7. List supervisor(s) at each job and at least two co-workers. Phone numbers should be included, if available.

Checking the applicant's work history can provide a lot of valuable information. How long does the applicant stay with a job? If a person is moving from job to job every few months, more than likely there is a problem. It may be because of theft, poor performance, or just someone who cannot find a job that suits them. In any case, you may be hiring a short-term employee.

Did the salary increase or decrease as the applicant changed jobs? If the salary decreased, the chances are good that the applicant was fired from his preceding employment. If the salary did not change significantly, you may have an applicant that is "trying to find herself."

If an applicant gives the reason for leaving a job as "personal," you are afforded a clear warning sign. Large companies and government agencies will frequently allow an employee to resign in lieu of termination. The employee is simply told that if she resigns, she will not have to tell future employers that she was terminated. Most large companies and government agencies have policies that allow only the confirmation of employment, the length of employment, and the salary. This offers the reporting party a degree of protection from lawsuits, if for some reason they cannot prove allegations about the ex-employee. This may be a practice you will want to utilize.

If possible, always contact the people who worked with the applicant. Co-workers can be a strong source of information. They talk more freely than supervisors and generally have more "inside information." Be sure and tell everyone you talk with that the information they give you will be kept confidential and then be sure you keep it that way. It will increase the information you receive.

All of this may seem like a lot of work, but if you are able to reach people right away, it may only take an hour or two. If you want to take it a couple of steps further, drive by the applicant's home. You should always try to drive by a couple of times—first, on a Friday night about

Embezzlement

9 P.M. Is there a loud party going on? Are cars parked on the lawn? Are people standing around in the front yard drinking beer? These may not be good reasons for not hiring someone, but at least you know a little more about the applicant. The second drive by should be on a Sunday afternoon about 3 P.M. Is the home kept up? Does the lawn need mowing? Do the cars need washing? Are young children running around without proper supervision? All of these questions may tell you how well the applicant will take care of your business.

In most states, arrest records are confidential and therefore unavailable for your review. However, cases pending in court are frequently a matter of public record that anyone can see (newspaper reporters check court records everyday!). Check with your local court clerk to find out what kind of information is available in your state.

Background checks may not always be necessary. If you live in a small town where everyone knows everyone else, your background check is already done. If you're hiring a friend or relative, you may also want to pass on the background check. However, hiring friends and relatives produces special dangers. By the nature of the relationship, friends and relatives can take advantage of you. It may be a matter of coming to work a few minutes late or going home a few minutes early. If you have other employees, this sets a terrible example and creates feelings of unfairness. The more your friend or relative takes advantage of the relationship, the more difficult it will be to control the other employees. Employees may feel they are being taken advantage of and start giving themselves little "bonuses" to make up for the extra work they have to do. If at all possible, AVOID HIRING FRIENDS AND RELATIVES—you will save money and relationships.

There are always friends and relatives that will try to get you to hire one of their friends. Even though this has the advantage of a recommendation by someone you trust, it can also have its disadvantages. Why does this person need someone else to find a job for him? Will this person take advantage of the fact that you have mutual friends? How hard will it be on you to let this person go, if he doesn't work out? If this employee is caught stealing, will the mutual friend put pressure on you not to prosecute? Will this employee feel that he is safe from prosecution? These are questions that only you can answer, but be alert to the pitfalls of hiring someone based on friendship.

Now that you have completed the background investigation, you have nothing to worry about—right? Wrong! Law enforcement agencies, the military, and defense contractors conduct extensive background investigations. But, they still end up with employees that slip through the cracks. Background investigations are critical in selecting employees, but quality can never be guaranteed.

Never tell an applicant that you are not going to hire them because they "didn't pass the background investigation." If you do, you will get angry phone calls from everyone you contacted during the investigation. The applicant will contact everyone she used as a reference wanting to know what terrible things they said. It is easier to tell an applicant that you are hiring a more qualified person. This is also better for business because you have not alienated any of the people you contacted.

Drug Use by Employees

Drug abuse is affecting every aspect of society. Some of the people you'd least expect are drug users and/or alcoholics. If an employee doesn't have a dependency when you hire him, don't relax your guard, for it doesn't take long to develop one. Catching the problem early is the secret. So what do you look for? It may be as simple as looking "hung over." Perhaps too many trips to the bathroom will tip you off. A subtle change in habits or attitudes may signal a problem. Energy swings are common of someone that may be using drugs at work. If you have an employee that looks tired and returns from the bathroom ready to tackle the world—you may have an employee that already has a serious problem. Drug users are often paranoid and careless with their work. Employees that suddenly start using absolutes in their conversations such as: NEVER, ALWAYS, ALL THE TIME, EVERYBODY, NO ONE and the like, may be showing the first signs of paranoia.

Usually, before the average person can detect drug abuse with any certainty, it's too late. So what should you do if you suspect drug abuse? First, talk to that person— he may have a perfectly good reason for what has been going on. If there is a drug problem, he is not likely to admit it. He probably cannot admit it to himself, yet alone his employer. If the problems remain after the talk, you may want to ask the employee if he has a dependency problem. Explain why you are concerned and detail the behavior you have observed. If the employee still denies a drug problem (they usually will), then request the employee to submit to a drug screen. Most hospitals and medical labs can provide an "employment screen" at a reasonable cost. If the person is using drugs, it is not likely he will show up for the screening or come back to work—except maybe to pick up the last check. What are your options if the employee refuses the drug screen? That depends upon the law in your state and you should check with a legal advisor.

Embezzlement

One of the most common drugs abused is cocaine. The following is a summary of the signs to look for in attempting to recognize the cocaine user:

Signs of Cocaine Abuse:

1. Cocaine paraphernalia—short straws, small spoons, small bindles of aluminum foil, and nose drops.
2. Dilated pupils.
3. Redness and soreness around the nose.
4. Rapid breathing.
5. Hyperactive, bizarre and/or paranoid behavior.
6. Itching of the skin.
7. Perspiration on the forehead or neck.
8. Extreme desire for liquid.
9. Redness in the eyes due to lack of sleep.
10. Needle marks on the inner arms.
11. Dark brown deposits on the tongue and teeth.
12. May wear sunglasses to shield eyes.

Symptoms will vary between individuals. Some will show almost all of the symptoms while others show few, if any. Additionally, other drugs, such as amphetamines, may create many of the same symptoms as cocaine. You should avoid accusing an employee of being under the influence of an illegal drug just because you recognize some of the symptoms. Even physicians and trained narcotics officers prefer to back up a diagnosis with lab tests. If you believe an employee is under the influence of an illegal drug, you should notify your local law enforcement agency. If you have a good reason to believe an employee is under the influence and don't do anything about it, you may be liable for any damage or injury caused or incurred by that employee, especially if it occurs during the operation of your business. As stated earlier, remember that when an employee becomes an addict, there is a paradigm shift—everything resets to zero. No matter how trusted, reliable, and dedicated the employee might have been before, such will not be the case now. The addict, by sheer necessity, operates under an entirely different set of rules that place the need for drugs (and money to buy the drugs) as the absolute first priority.

How Employees Commit Thefts

Employee theft can be as simple as taking stamps or as involved as accepting kickbacks from vendors and suppliers. The most common thefts are pilferage of inventory and the "short ring up."

Pilferage

Pilferage, the theft of inventory, is believed to amount to 30-75% of all shrinkage of inventory, and creates a loss of 5-10 billion dollars annually. Believe it or not, pilferage is often encouraged by business owners. It begins with some inventory being damaged and the owner allowing the employee to buy it at reduced cost. Frequently, employees are allowed to purchase advertising or display items below their associated costs. If your employees select the items to be displayed, they may be making decisions based upon their future needs. At some point, they realize that all they have to do is set something they want in the window for a minute or two, then "legally" buy it as a "display item." If an employee is given the authority to declare damaged or otherwise non-working products as non-sellable and designate them to be sold at a lesser price for parts, problems may occur. In many cases, employees will purposely damage the item or render it non-functional just so they can purchase it at the lesser price. In one case, the employee did little more than unplug a power transformer necessary for operation. Once he purchased the item, he merely reconnected the transformer and ended up with a brand new piece of expensive electronics for which he paid only a fraction of its actual cost.

Business owners have, in some instances, allowed their employees to make personal purchases directly from their wholesale source. If this is not closely monitored, employees soon learn that they can order the item for the store at wholesale price, then simply forget to pay. The business owner that does not closely watch inventory ends up giving more gifts to his or her employees than is realized. Since the bill that will arrive will be from a usual supplier, it is usually paid by the bookkeeper. While on the subject of bills, be careful that a less than scrupulous bookkeeper isn't making up generic bills and then paying them—the bookkeeper is the ultimate payee, of course.

Some companies allow their employees to remove equipment or products from the business for their temporary use. It may be the removal of a computer, an adding machine, a sound system, an electronic typewriter, a video cassette recorder (or a million other things). If you set up a tight check-out system that you monitor, you may not have a problem. However, if there is no check-out system you are asking for trouble. You have given your employees license to load your property into their car and leave with it and you may never know who took it or whether or not it ever came back. When something is discovered missing, it will be unknown whether you've been burglarized or become a victim of embezzlement.

Embezzlement

Another scam involves an accomplice who returns an undamaged, functioning product purchased earlier with a claim that it is either damaged or non-functional. A receipt is written for the item and the "customer" is given back the money paid. The clerk then sells the item for cash. Since products to be returned to the manufacturer do not appear on the daily receipts, neither the return nor the cash sale will be recorded. Therefore, there is no cash nor inventory discovered missing. This scam could be committed repeatedly if it is spaced out over a period of time and if the business did not institute some policy regarding managerial approval of returned goods.

Employees can also be influenced by friends to steal. This is especially true of young employees. It may begin by allowing a friend to take something with the promise that it will be paid for later. When later never comes and the employee doesn't get caught, a cycle begins. If employees have friends who come into your business, be sure that some kind of pattern is not being set. Do the individuals appear to be making purchases? If so, are the purchases being rung up properly? Do the employee and her friend stop talking when you get close? Do they look around to see where you are and if you are watching them?

If you see an employee's friend buying something, take note of what it is. Then, as soon as possible, check the register to make sure the employee didn't short ring the item. Employees will sometimes charge a friend far less that they should for a particular item. Another method that is used is when a friend or conspirator is given more change than they have coming. Of course, some employees will not make the fact that they know the person obvious if they are planning a theft. Be alert if a customer seems to wait for a particular employee to help him, especially if they belong to the same age group.

Sometimes employees will just look the other way while a friend shoplifts. They may even try to distract you or other employees while their partner makes off with the merchandise. If someone suddenly leaves the business after an employee has taken you aside, be alert for future patterns.

It is a good idea to check the trash cans occasionally. This should be done just before they are taken out for the night. Employees will sometimes take the property out in trash cans, with plans to retrieve it from the trash bin at a later time. In one case, a shipping and receiving clerk would take property from the package delivery van while it was parked outside the receiving dock. He would step inside the van while the driver was inside the business obtaining signatures for the delivery and grab any packages destined for another business—a gun shop. He would then walk nonchalantly out to his vehicle, taking a split second out to drop the property into a large trash dumpster. He would get into his car, as if retrieving something, and then go back to work. If the supervisor ever suspected the employee of theft and searched his car, he would find nothing. At the end of the day, the employee would attempt to be the last one leaving. When he did, he would retrieve the item from dumpster and place it in his car. His actions were photographed by the police, who positioned a photographer, armed with a high-power telephoto lens, in a tree some 50 yards away. The police did not suspect the shipping clerk, but they suspected that the thefts might be taking place at that location. Upon his arrest, the reason behind his repeated thefts was found in a small glass vial—cocaine. It was estimated that the guns he took from the delivery van totaled nearly $20,000. And, it put a lot of stolen handguns on the street, primarily in the hands of cocaine dealers.

Receiving clerks, security personnel, and others have also been known to obtain duplicate keys to storage areas and to the store itself. They simply return after hours to commit the crime of burglary.

Embezzlement of Cash

To battle embezzlement of cash, the payment counter is the most likely location to watch. In fact, many businesses place pinhole video cameras directly over the cash register. These cameras are hooked to a time-lapse video recorder and may be run during the entire time the store is open, or as needed when suspicions of a particular employee arise. They are excellent evidence.

The "short ring up" involves punching in (ringing up on the cash register) a lesser amount than that of the purchase. The employee then keeps the difference. This method would still allow the cash register drawer and tape to balance. In some cases, the clerk simply fails to ring up the item purchased. Rather, they hit the key or key combinations that allow the drawer to be opened. Once the cash is accepted, change is given. The difference between the change and the amount paid is taken by the clerk. Again, since no sale was entered, the missing cash would not show up on the tape nor in the drawer.

Employees may purposely fail to mark items that are on sale. They sell the product for the regular price and then pocket the difference. Or, if the item is marked for sale, they may charge the regular price hoping

Embezzlement

that the consumer won't notice. It would appear as an honest mistake. An employee may accept cash for payment and then enter a *negative amount* into the register of the same amount as the purchase. They then keep the cash paid to them. In this way, the cash drawer totals and the internal tape will agree. Although the negative entry will be on the tape, it may take hours to locate it. Since the drawer and the tape agree, short of a witness, there would be no reason to suspect anything was amiss.

At the point of sale, employees may accept the cash payment and fail to put the cash into the register. For example, a common technique is when the customer pays with a single bill (such as a ten, twenty, or fifty). The employee accepts the bill, acquires the necessary change from the register, and delivers it to the customer. One step gets missed, however—the employee never puts the single bill, offered as payment, into the register. This is commonly called "palming." The employee, in the process of making change, will either wad or fold the bill inside her palm, holding it in with the thumb. Shortly after the transaction is completed, the employee merely puts the hand in a pocket, as if looking for something, and releases the bill. If they're good at this, it is even difficult to catch on video.

Employee Receipts

If your business employs persons to make deliveries and an individual merely submits fuel and repair bills without any further verification, beware! There have been many cases where a shop/gas station employee has worked with truck drivers and delivery persons in scams such as this, splitting the difference.

Where expense accounts for travel, lodging, and meals are concerned, require the employee submit printed receipts from the register, not handwritten receipts. Restaurants, in cities that cater to the convention crowds, routinely have bundles of 10 or more blank receipts that are given to customers requesting a receipt for their meal. Recognizing that sequentially numbered receipts would create well-founded suspicion, they generally ensure that they are from different number series books. Persons who travel a great deal for the company, and who are required to provide receipts, generally collect them religiously. The ones that do not have dates or amounts are "keepers" to be used later to collect reimbursement.

If receipts are not verifiable, it's a red flag. Most receipts from electronic cash registers include the name of the business, the address, the date, the time, and the initials of the cashier. Have a rule which states, "Receipts that cannot be verified will not be paid."

Embezzlement

Manipulation of Accounts

Manipulation of accounts is another arena of embezzlement. Persons who have access to the accounting procedures know that if they manipulate numerous accounts, only a complete audit will uncover the losses and the pattern. An examination of any single account is likely to identify an aberration, but just as likely to be identified as a clerical error.

In larger companies, payroll clerks have been known to fail to remove the names of persons who have left the company. Checks are made out to the employee(s) and the payroll clerk (or someone else along the line of pay) merely endorses the check and cashes it. In other instances, a person having direct contact with the payroll will create a fictitious employee. When the checks are printed, this individual merely endorses the check through a process known as "forgery" and cashes it.

Some employees, who have access to or are responsible for writing checks are not afraid to write checks to themselves. If they're uneasy about doing that, they simply write a check to a friend and then split the cash.

There was a case where a computer programmer who provided services to a number of financial institutions decided to see what his expertise would allow him to do. He set up a computer debiting program that took a fraction of a single penny from each transaction in hundreds of accounts and deposited the money (electronically) into another account with a fictitious name. The tiny amount of money involved from each individual account was not enough to raise a red flag for the auditors to see. When examined cumulatively, however, they would have seen that literally millions of dollars were being transferred to a single account set up by the crook. One day he decided enough was enough, and had his funds transferred to another account in Rio De Janeiro, Brazil. There, he lived the good life. Unfortunately, his ego got the best of him. He couldn't stand the fact that he so brilliantly masterminded what was probably the greatest fraud in the history of the world. He returned to the United States and began bragging about what he had done. He was arrested, tried, and jailed. So much for Rio.

Records

To reduce the likelihood of unauthorized disbursements, make sure that you have a rule which prohibits the removal of certain records from the premises. Once out of the premises, records are much more easily altered or destroyed. Any employee that would violate such a

Embezzlement

rule is an employee who has stolen or is stealing from you. In many embezzlement investigations, the paper trail is all you have. If the paper trail is in the possession of the perpetrator, the perpetrator has all the cards.

Any business owner should have a checks-and-balances system in place to reduce the likelihood of internal theft or fraud and to identify it if it occurs. It is important that records be audited routinely—some scheduled, some not. The cost of an internal audit is a fraction of what can be lost through embezzlement if audits are not carried out.

Do not assume that an employee who has been performing a clerical task without a problem for ten years won't have a major problem in the eleventh year. If you catch it early, you may save your business.

Periodically check account balances, look over the disbursements, and never, *never*, sign blank checks for someone else to complete. If you send a signal to your employees that you are on top of things, they will be hesitant to steal from you. If you run a loose organization, however, you invite theft. Remember—the likelihood of theft is in direct proportion to the employee's belief that they might get caught. Make them think they'll be caught—every time.

The False Report of Robbery

Embezzlement of company monies or products may be accomplished by way of a false report of an armed robbery. Employees have greater access to the money and to the products than anyone on the outside. If the employee is a good enough actor or actress and can keep the story given to the police on-line, it will probably be investigated as a valid armed robbery rather than an internal theft. However, there are some indicators that, when viewed in their totality, strongly suggest internal theft, rather than robbery.

Obviously, employees who pilfer small amounts of cash will give little thought to when they take the money, because they figure it will be written off as a clerical error. If they're going to take a large amount of money and attempt to cover it by reporting a false robbery, they will specifically choose a time or period when they know the cash on hand is high. Exactly when this time period *is* will depend upon the type of business. Some businesses will have more cash on hand around the first and the fifteenth of the month, when assistance checks are issued. For a convenience market, weekends (especially three-day weekends with holidays attached) are periods when large amounts of cash are on hand. Certainly, many crooks know this, as well. This is why you cannot use any single factor as justification or evidence to believe that a reported armed robbery is untrue. However, this, in conjunction with the other factors that are to follow, may be evidence enough.

If the employee is going to rip off the employer and hide it behind a false report of an armed robbery, it will generally be done when that employee is working alone. The employee may choose a time when the second clerk has left to pick up a meal or make a delivery. If the employee is working alone, there is ample opportunity. Again, this in and of itself, is not conclusive evidence of anything, as crooks, too, have been known to watch businesses and strike when staffing is at its lowest.

Most persons that are robbed (even if it has happened to them before) are extremely upset, even long after the event is over. This is especially true of females (who may also fear abduction and sexual assault). The trauma is greatly increased if the perpetrator is loud, profane and commanding, and threatens to take the life of the employee. The trauma level shoots skyward if a weapon is displayed and pointed at the employee. When the perpetrator wears a mask, employees tend to focus on the eyes and often describe them as "psychotic" or "crazy looking," which is just the appearance the perpetrator wishes to project. If the employee you encounter (after being summoned to the scene) is calm, cool, and collected, or is even laughing at the whole event, it should send up a red flag.

Additionally, most employees placed in such a traumatic situation are marginal witnesses. Fear destroys our perceptual skills and clouds our memories. Young girls working alone at night in a convenience market are not generally the epitome of calm under pressure. Therefore, it is expected that their ability to recall the events should be somewhat limited. Where they give themselves away is when they are "too good of a witness." What they have done is pre-selected a description to give the police. It may be loosely based on someone they know. Often, however, by the time they put their plan into action, their nervousness prevents them from recalling all of the information they so carefully assembled. This is why either you or the law enforcement officer should get the employee to write down the events and the description of the suspect as soon as possible. Take the document and retain it. As the story is repeated during the investigation, it will be quite easy to see if things change—and often they do.

In some cases, the employee will key on questions asked by the officer and use them as a "feed" to decide what the answers should be. For example, if the officer asks if the suspect had an earring, the employee may jump too quickly and answer, "Yes, that's right!" as if suddenly remembering. What the employee has forgotten is that he's already told the officer that the suspect wore a heavy knit ski cap that covered his entire face. If the officer asks (and emphasizes its importance) if the suspect had a gold tooth in front, the "seed is planted." About now, the employee is thinking that the police must be looking for a robber with a gold tooth already and may see this as an opportunity to latch onto an already established string of crimes. The employee will answer in the affirmative. When asked if the crook had a tattoo of a peacock on the inner aspect of his right forearm the clerk may jump at that, too—not realizing that he'd already described the perpetrator as wearing a long sleeve shirt and having never rolled up the sleeves. No matter what the officer suggests, be it a gimp, a limp, a lisp, an unusual accent, or a polite nature, this employee will answer in the affirmative. The reality is, few witnesses, unless they were police officers, are ever this good.

On the opposite side, the employee may give a grossly vague description of the perpetrator. Although having had the opportunity to view the individual, the employee can remember little or nothing. This is generally because the rehearsed description is being drowned by the fear of not being convincing enough and not being believed.

When confronted with even a minor inconsistency, the embezzling employee will become defensive. Most legitimate victims become frustrated with themselves at such times for their poor memory. Remember, they want this guy caught. The untruthful employee, however, is hypersensitive to any suggestion that they have given false information (even though it was only suggested they were mistaken or confused) and will jump back at the questioner. At some point, the employee may defiantly say, "Don't you believe me? Do you think I did this!?" And, believing that one should be extremely upset at such a scenario, the employee may flatly refuse to answer any further questions and may verbally attack the officer, suggesting he or she is unqualified to handle the case investigation. The employee may suggest that the employer is using the event as an excuse to fire him, something he (the owner) has wanted to do all along, generally linking it to ethnicity, perceived favoritism, a prior grievance, etc. If the employee suggests or asks for the presence of counsel to represent him, you can be reasonably certain that this theft was a theft from the inside. Throw another flag on the pile.

The employee may also get creative and injure himself slightly (usually a small cut of some sort) to add realism to his story. Some have been

Embezzlement

known to have the cut bandaged prior to the arrival of police. Now there's something that most robbery victims would do right away! Most robbery victims hide until police show up! Most employees who are injured by an unknown assailant are fearful of everything from Hepatitis B to HIV and will want medical treatment. If the employee declines medical treatment, select another red flag. Or the employee may overturn a table or display to coincide with the reported struggle or the actions committed by the perpetrator in his fits of anger. In one case, a table was observed laying on its side, reported by the employee as kicked over by the suspect. And while the display items were on the floor in disarray (as would be expected), the two potted plants that were also on the table were found sitting against the wall without a single granule of soil missing. Now that's some magical kick! In another case, when investigators attempted to recreate the action of the suspect, they were unable to cause the table to land on its edge, as it was found. In every case, the table turned all the way over, legs up. So then, the police would have to believe that the suspect actually took the time to set the table on its side for some unfathomable reason. Choose another red flag.

In one case, a delivery driver that took in cash payment reported that he was robbed by three men. The red flag in his story was his explanation that after the robbery he drove home to call the police. Yet, he had police meet him at a nearby store—one of many locations he passed en route home that had pay phones. He went on to describe his attackers in exacting detail, including the tattoos he saw on the one suspect whom he had just said was wearing a trench coat. No matter what characteristic the officer suggested, this particular "victim" said "Yes." As a ruse, the officer went back over the descriptions (to verify), adding information the "victim" had never offered and even changing the ethnicity of one of the assailants from African-American to Asian. And, right on cue, the "victim" replied, "Yeah, that's right." The money was found at his home and both he and his wife were arrested.

Robbery suspects seldom work any harder than they must. When cleaning out a cash register, they take bills and, in a very few cases, fifty-cent pieces or quarters. They don't mess with loose dimes, or nickels, and certainly not pennies. Rolled coins, although easy to grab, are heavy and cumbersome—so they're usually left behind. Checks are not taken because they are too easily traced back to the source, especially by the stamp "For Deposit Only," which is followed by the name of the business robbed. If your employee's crime contradicts these time-tested, known methods of operation, add yet another red flag to the pile.

Embezzlement

Robbers, when they do wear masks, do so to either prevent themselves from later being identified in a line-up, or because they do not wish to be recognized. In the latter case, the robber may be a present or ex-employee.

Take a close look at "the way the crime was committed." Most robbers want to get in, get the goods, and get out as quickly as possible. They make every effort to control the clerk so as to reduce the likelihood of the clerk arming himself or setting off some alarm. If your clerk was directed by the perpetrator to go to the back of the store, get the keys to unlock the safe, yet was not followed and did not use this opportunity to call 911 from a phone in the rear of the store or hit a door in order to get out of there, then you can grab another red flag. Crooks do not send store personnel anywhere out of their sight—ever.

You now have a pile of red flags, all suggesting that this robbery was in fact, an embezzlement. In most cases, the money will still be somewhere on the premises. Initiate a search. The money could literally be anywhere—in a locker, in the clerk's purse, or car.

Do you have a clause in your contract with employees that allows you to search lockers? Personal vehicles? If not, so long as you're not searching at the request of the police, you may still be on legal grounds. Rear storerooms are great hiding places, as well as in the rest rooms within fixtures. One male clerk placed the cash in a plastic zip-lock baggy and put it in the toilet tank. A female clerk placed the money in the paper towel dispenser. Don't forget the trash enclosure and the dumpster—they are common locations, because they allow for easy retrieval later, under the guise of emptying the trash at closing time.

Keep in mind that an innocent employee reporting a legitimate robbery might be a bit hurt that you are asking to search personal areas like lockers and cars. They seldom will object, however, because they have nothing to hide and wish greatly to clear up any suspicion—they don't want to lose their job. The guilty employee, however, is likely to scream out "I quit!" and then attempt to leave. Calm the employee down and ask him to remain at the scene (and there's a reason for doing so).

It is important that you "strike when the iron's hot." If the employee is allowed to leave and return home, it provides time for reflection, calm, and preparation for what the next day may bring. Ideally, the police would confront the clerk with all of the inconsistencies (literally dump the red flags on the employee's lap) right away, while emotions are still running high. By showing empathy and giving the employee a way to save face, you may just get a confession and a recovery of the money. Suggest that the employee has a drug habit that they can't control, or is involved in a relationship with someone who has such a problem and who is forcing the theft to occur. It is important to tell the employee that you do not believe that he or she is a "bad person," but rather, that they just became a victim of needs they couldn't meet through legitimate means. Remember, empathy doesn't mean you agree—just that you understand (even though you may not). Make every attempt to get the employee to write out a formal confession (even if you intend not to prosecute) to protect you from civil litigation for wrongful termination.

In some cases, you may opt to dismiss the employee rather than seek prosecution (although in many states, the false report is a criminal violation over which you, as the business owner, have no control). Certainly, there is a great value in determining, at the time it is given, whether or not the report is false. The investigating agency was saved countless hours of follow-up investigation showing line-ups, looking through mug books, etc.

Other Considerations

If an employee has the authority to purchase supplies or design displays, suppliers will occasionally bid for their good will. They may offer small gifts or give them cash under the table. Be especially alert if suppliers and key employees seem to be overly friendly. Ask yourself, is one brand of product getting more counter space than others? Does the employee push one product over another? If so, is it for a good reason? What is my cost of this product as compared to others? Are my customers being better served by this product?

If your employees work overtime on a regular basis, try not to cut back on it too suddenly. People who are accustomed to working overtime will often make their purchases based upon it. If overtime is suddenly cut, they often find themselves in trouble. This may encourage sabotage to create overtime opportunities.

Dishonest employees have many ways of hurting your business. The *abuse of sick time* or *on the job injuries* (fake) are two of the more common ways. The best way to protect yourself is knowledge. A basic course in personnel management is almost essential for employers. Although you can always fire someone for theft, do you know your rights and responsibilities with other issues? There are state commissions that

Embezzlement

investigate complaints filed by employees everyday. By knowing the laws in your state, you can protect your business against the disgruntled employee as well as the thief.

Recognizing the Embezzler

There is no magic technique allowing you, as a business owner, to recognize embezzlers by looking at them. They do not glow in the dark and they are careful, in most cases, not to give themselves away. There are some generic signs, however.

The employees who fail to take their vacation or desire to work on their days off may not be the highly motivated, dedicated individuals you think them to be. The reason they make it to work every day is to cover their behinds. They do not wish to risk the chance that, in their absence, someone will stumble over their activities.

The employee who disappears suddenly should definitely come under scrutiny. Few people just pick up and leave and are never seen again, unless there's a reason. If you have an employee who disappears off the face of the earth, you'd better start looking at accounts and at inventory. Chances are that you've got a major problem in one or both of those areas that you weren't aware of. This is especially tell-tale if the employee makes a hasty exit at the mention of an audit, inspection, or immediately after the discovery of a problem with the books or inventory.

Cast a cautious eye upon any employee who acquires sudden prosperity. An employee who is living month to month, trying to make ends meet and suddenly is driving a new car, purchasing property, etc., should raise an eyebrow. Unless their spouse got a major promotion or a rich relative died and remembered them in the will, there's a good chance that "sudden prosperity" came as a result of stealing your money and/or inventory.

If you have an employee whose social activities are grossly inconsistent with their income, view this as a sign that something outside of the law may be occurring. Granted, it may not be theft, but it may be the sale of narcotics (which leads to theft).

Finally, if you have an employee who seems constantly nervous or worried without apparent reason, consider this to be a red flag. This is especially true if their emotional upheaval increases when the boss is around, when the books are being examined, or an audit is planned. You may find that such an individual becomes both irate and defensive if someone gets into their desk or looks at their record keeping.

Embezzlement

Arresting The Problem

It is not unusual that a business owner first discovers employee theft when they accidentally catch someone stealing from them. So what should you do? Call the police! Although you may or may not want the hassles of going to court or the embarrassment of an employee being taken away in handcuffs—call the police anyway! It is the best way to protect yourself. A police report can help protect you against civil suits, labor complaints, and even criminal complaints. It is not unusual for someone that has been fired to make up stories about their boss. A sympathetic friend or relative hears the story and insists that the ex-employee file a complaint against you. The ex-employee is not going to admit to a relative or friend that they have lied, so a false complaint that you will have to defend against is filed. Some of these complaints can be serious. The more serious the complaint, the more sympathy they will receive from friends. They may also believe that you will drop the charges, if they promise to drop the "charges" against you. Police officers are frequently targets of these types of charges and will be suspicious of employees who try to bring charges after they have been arrested.

If the theft is discovered accidentally, take the employee into a private room, if possible. Tell the employee that you wish to discuss a problem with her. Excuse yourself from the room, then call the police from another location. While you are waiting for the police, talk to the employee (keep it friendly). If you conduct the interview in your office, avoid a seating arrangement where there is a table or desk between you and the employee. Sit in a chair facing the employee and move in close. This allows you to create tension to get the truth and to reduce it as a reward for telling the truth. When she begins to tell the truth, slowly move your chair back. If she begins to lie again, move in closer. If you can make the employee feel guilty, rather than defensive, you are much more likely to get a confession. You may want to ask, "Haven't I treated you well?" or, "Why would you do this to me?" or even, "What's going on in your life that would make you have to steal from me?"

Why do you want a confession? A confession to you will pave the way for a confession to the police—it will be easier to say the second time. Tell the officer (in front of the employee) that the employee has admitted to the theft. The employee will find it difficult to call you a liar. For some reason, most people find it easier to confess to a crime than to call someone a liar to their face. The police officer will hear the confession

Embezzlement

and can back you up in court, if necessary. If at all possible, avoid the word "confess." Instead, "admitted to" or "told the truth" are far better ways of saying it in front of the employee. The word "confession" brings to mind the word "jail," and the employee may decide not to talk. A confession (especially to the officer) will support your observations and help keep you out of court.

After the police have completed their investigation and left your business, start making thorough notes. Most police officers go from call to call without enough time to write a thorough report. The officer will get just enough information to get the case into court. The case may not go to court for six months and a lot of details will be forgotten. If you keep good notes, you will be sure to have a good case.

What if you just suspect employee theft? First, look for a pattern. What is being taken? Is there a certain day of the week that it is occurring? Is it happening at the first of the month—when rent is due? A certain time of the day? Which employees are working when it happens? If you can spot a pattern, it will make it a lot easier to catch the thief. If there doesn't seem to be a pattern, you may want to talk individually with each employee. Tell them there are thefts occurring and you would like them to keep their eyes open. You might want to tell them that the losses are hurting the business enough that you may have to lay someone off. This may cause an employee to come forward and name the thief. It may even be that the thief will stop because he doesn't want to lose his job or cause someone else to get fired. If the thefts stop right after you have had this talk, you can bet that an employee was involved. The chances are good that another employee will name the thief. If this happens, get enough information so you can catch the person red-handed. Try to find out if they were witness to the theft or just heard about it. If the person heard about it from another employee, find out which one.

If your employees knew about the thefts and remained quiet, you have a more serious problem. If you fire someone just on information from an employee, it is not likely that an employee will come forward in the future. A lot of people have problems with being a "snitch," so people who pass along information will often insist on anonymity. When possible keep their names out of it. You may need more information in the future. If it is a serious theft and the employee was a witness to it, call in the police for advice. Many times, if you have a good witness, the police will make an arrest and conduct a thorough investigation. Keep in mind that the thief may identify an innocent employee to remove suspicion from himself. Whatever the case, tell all employees from the start of their employment that they have an obligation to report all thefts.

It is always a good idea to have a witness whenever possible. It will not only help when you get to court, but will also help prevent false allegations, as discussed earlier.

There are a number of security devices on the market that can be useful in preventing theft. Most large companies use these devices as much to prevent employee theft as anything else. However, before you invest in any hardware, ask to test it for several weeks. Check to see how easy it is to use and if it will work for your application. Many of these devices are overrated, so check them out carefully. The following chapters will discuss security systems in more detail.

When should you give an employee a break and not have him arrested? The safe answer is never. Having an employee arrested is not just to protect yourself, it is sending a clear message to other employees of just how serious you consider the matter. However, it is your business. Let your heart (and don't forget your pocketbook) be your guide.

Once an employee has been fired, it is not unusual for the business to be the victim of a bomb threat. Bomb threats can be unsettling, but they are seldom anything more than just a threat. Does the voice sound familiar? Does the caller seem to have information only an employee would have? A disgruntled ex-employee will certainly keep making the threats if he sees he can disrupt your business anytime he wants. However, always close the business if a suspicious item is found.

Naturally, the only course of action is to notify the police. The decision of whether or not to clear out the premises or close for the day will most likely have to made by you, as most police agencies will not take the responsibility.

Case Study: The Anatomy of an Embezzlement

An attorney with a law firm checks the balance of a trust fund being held by a local bank. An account which should have contained some fifteen thousand dollars was only two or three hundred dollars strong. He then checks with his secretary (who made all of the deposits for the client) who, outwardly, is as shocked as he. She suggests that the account number must have been communicated incorrectly. When the attorney goes back into his office to make a second call to the bank, the secretary remembers she has a doctor's appointment and hurriedly leaves the building. She leaves her purse behind. Now that's unusual.

Embezzlement

The preliminary investigation by the law firm reveals that they had only found the tip of the iceberg. There were several accounts that had been cleaned out. They began to suspect that the secretary, whose job it was to make deposits and reconcile accounts, was responsible for a massive embezzlement.

At first, the partners refused to believe that she could possibly be responsible. She was a trusted employee who had been with the firm for many years. She was the epitome of efficiency. If you needed to know something or needed something done, she could do it for you—not only quickly, but very well. Then they discovered that the ledger book was missing. This book, which was the documentation of all cash inflow and disbursements, was never to leave the building. Then, they remembered that the secretary's work had slipped somewhat since she married a "less than reputable" man.

They called the bank to see if today's deposits were made. None of the business tellers recalled seeing the secretary in the bank on this day. "I have the deposit receipts!" exclaimed one of the attorneys. Yes, they were slips of paper. No, they were not like any bank receipt anyone in the room had ever seen. A quick look through the secretary's desk revealed the source of the receipts—a rubber stamp and blank forms.

The senior partner then admitted something. He had been signing entire blocks of blank checks for this secretary. It was too much trouble to sign one at a time, and besides, the employee was trustworthy (she wrote these checks to herself and her husband). He admitted that he'd never felt it necessary to verify any of the checks she had written, deposits she had made, or information she had provided.

A police report charging the secretary with embezzlement was written. The true amount of loss, however, would not be determined for several days. It would take several accountants several days to straighten everything out.

A search warrant was issued for the home of the secretary where law enforcement officials hoped to find the ledger book. There was a dim hope on their part that the stolen money might be in a shoe box, someplace. It was an unrealistic hope.

The search warrant was executed and the ledger book was found, along with other documents that also were "never to leave the offices." The secretary was not at home when the search was first initiated, but arrived home when it was in progress. While most people (innocent people) who arrived home to find police vehicles in their driveway would jump out and hurriedly run inside to see what had happened, this particular person did not. She backed out of the driveway, then accelerated away at high speed. After an 85 M.P.H. pursuit, she pulled to the side of the road. She was then taken into custody.

A subsequent search of the contents of her purse revealed a vial of cocaine, single-edged razor blades, a mirror, and a short straw. It seemed our trustworthy, tenured, super-secretary also had a cocaine habit. Remember what was said about addictions: When someone is addicted, there is a paradigm shift—everything resets to zero and all priorities change. By the time the audit was complete, the loss had tripled. Where was the cash? It went up her and her husband's noses. The law firm made the classic mistakes. True, they may have been too busy to monitor everything and verify some things, but then again, could they afford not to?

Basic Rules of Conduct

1. Be careful about establishing any sort of policy where damaged merchandise is either given to, or allowed to be purchased (at a greatly reduced price) by store employees. Dishonest employees have been known to purposely break items simply to get the discount. The scope of this runs from food items to electronics—there isn't anything for which some people won't commit fraud.

2. Always maintain a zero-tolerance policy with regard to losses and short register counts. Even if you have an "acceptable" loss parameter, do not communicate this to employees.

3. Prepare receiving reports immediately upon receiving the shipment. Failure to do so may result in the opportunity for theft or errors in record-keeping.

4. Always be alert for over-rings and require employees to contact you in the event that there is an over-ring. In the event an employee fails to do so, take quick and stern action. This shows other employees that you are serious about this policy.

5. When items are returned and a cash refund is requested, never allow the employee who made the original sale to handle the refund. The "customer" and the employee could be working the refund scam together. Require that someone else verify the receipt and handle the transaction.

Embezzlement

6. Be careful about which employees you allow to mark (price) items for sale in your store. Only allow those you have authorized to do so. Also, mark items by machine or stamp—do not allow the pricing of items to be made in handwriting.

7. Prosecute shoplifters and employees who steal from you. If you allow restitution and then drop the charges, you are inviting theft from others.

8. Set high standards for all of your employees and be the role model they need.

9. Ensure that you do not inadvertently set a double standard of moral and ethical conduct. If you receive an over-shipment, make sure it is sent back or otherwise reconciled. If someone overpays on an item, make certain that they are given the money back. What's required of employees should be no different than what is required of managers. If a manager violates a policy, it should be well known that he or she received the same level of punishment that a lower-level employee would receive.

10. Treat your employees well. Employees who are treated fairly (where benefits, pay, incentives, etc., are concerned) are less likely to steal from you. This is not, however, a guarantee.

11. Show your employees that you truly care about them. When an employee is making that decision to steal or not to steal, hopefully the employee will think about the time you offered praise in front of others or gave a day off when it was needed. For most people, it is difficult to steal from someone they respect.

12. Set reasonable goals and expectations for your employees and ensure that they have both the time and the necessary tools to meet those expectations. Failure to do so will almost always result in low morale and will make it easier for them to justify stealing from you.

13. Items taken off the shelf for use in the store, whether it be a pair of scissors or floor cleaner, should be marked on a special listing. This shows the employee that you expect a 100% accounting for everything (even though that is unrealistic).

14. Inspect the trash disposal bins. Employees have been known to place the item they wish to embezzle in the trash, carry it out to the dumpster, and place it in there for later retrieval. Often, these employees are the ones who always seem to be the last to leave (allows for retrieval of the item without being seen).

15. Conduct unexpected audits. The secretary who worked for a law firm and cleaned out a number of trust accounts—that embezzlement could have been caught early on if someone had conducted an audit of the books.

16. Be very careful about to whom you issue keys. If allowed out of the business, copies can be easily made. Make sure all padlocks are re-locked on the hasp after doors are opened for the day's business. This will prevent lock changing by employees planning to return later to commit theft.

17. Depending upon the type of business you have, it may be wise to prohibit the bringing of lunch boxes, purses, etc., into the work areas—these are places to hide items to be stolen.

18. You may want to have the company mail delivered to a Post Office Box rather than coming directly to the business. This will give you or your designee sole access to incoming checks which could, in the hands of an unethical employee, be used to juggle the books to cover prior embezzlements, or be converted to cash.

Chapter Three

Shoplifting—Anything But Petty

Introduction

In one state it may be called petty larceny while another may term it petty theft. *The reality is that shoplifting crime is anything but petty.*

Shoplifting crimes result in over $26 billion in annual losses, increasing consumer costs some five to seven percent. As a result, many of America's small businesses are finding it increasingly more difficult to sell their products at a reasonable profit while remaining competitive in the marketplace. Left unchecked, shoplifting crime has the power to drive any small business *out* of business.

Some see shoplifting as a "major retail chain store" problem. And while major retailers do experience significant problems with shoplifting, they are in a much better position to handle it. Most have asset protection/loss prevention units staffed by individuals trained to prevent thefts, identify and apprehend shoplifters, interview suspects and testify in court when called. When prevention and apprehension fails, these larger stores have the financial base to more easily absorb the losses. The small business seldom has any of these advantages.

The small business is generally minimally staffed, with only one or two clerks on duty at any given time. The clerk's time is divided between helping customers, stocking shelves, and taking care of inventory in the back room. There is very little time available to watch for customers who "look like shoplifters." Even if they had the time, most clerks have not been trained to recognize the techniques used by shoplifters and generally do not become aware that a theft has taken place until finding an empty box on the shelf or a garment-less hanger on the rack. Few have been trained in how to make arrests nor the personal and civil liability associated with doing so.

Thieves bank on a lack of preparation and take advantage of it. They discuss it with their cell mates and friends, which may cause your business to be marked as an easy target.

The goal of this chapter is to provide the reader with a profile of the shoplifter and their characteristic demeanor. Prevention of shoplifting, which will be discussed at length, is the optimal solution.

The Profile of a Shoplifter

This title may be somewhat of a misnomer, as there really is no true profile of a person who shoplifts. Experience has proven that the shoplifter can be either male or female, young or old, rich or poor. A person's appearance (with some exceptions) is probably the worst indicator of potential criminal activity.

Studies in the 1960's identified that shoplifters were most often women. This was understandable, in that few women worked outside the home and therefore took on the primary responsibility, as wives and mothers, to shop for the family. Generally, they shopped in department stores having security and when caught shoplifting, almost always pleaded guilty to avoid the embarrassment of trial. Today however, the gender of the shoplifter is not nearly so one-sided, with more and more men crossing the threshold of retail stores shopping for both themselves and their families.

What is the motive behind shoplifting? The only answer is that it varies with individual. Some steal things that they feel they need and cannot afford while others steal for the simple exhilaration that a successful theft brings to their otherwise boring lives. There are also those who steal items they deem too embarrassing to purchase, such as a kid pocketing condoms or even an adult, in need of over-the-counter herpes medications. Some shoplifters justify their thefts because of poor service, overcharges, or prior purchases of poor quality items.

The professional shoplifter steals for a living, usually to support an addiction. The goods may be put to personal use or sold to pay for rent, groceries, and whatever substance they are addicted to. Jail has provided them with an excellent education. They have learned the best ways to steal and how to manipulate the situation to their advantage, if caught. Surveillance cameras and security mirrors mean little to these people—they simply wait for the opportune moment and take what they want. It is this category of shoplifter that many retailers identify as the hardest to catch.

Demeanor Is The Key

An experienced security officer can generally spot a shoplifter the minute they enter the store. Experience has shown that shoplifters, through their demeanor, generally give themselves away.

A Focus On Personnel

The first indicator that a person entering your store may be intent upon stealing is a focus on personnel, rather than the product. Most persons entering a store to shop legitimately look first at the merchandise to see how the store is arranged to determine where the items of interest to them are situated. A shoplifter, on the other hand, is much more likely to focus first on personnel in order to determine how many employees are on duty and where they are.

Arrive, Leave, and Return

A shoplifter will generally walk directly to the area within the store where the property they wish to steal is displayed. Once there, the shoplifter will determine whether or not the item is present and, based upon size and packaging, how best to conceal (or *if* to conceal) the item. Generally, the shoplifter will not take the item at this time, but rather, will leave and go to another part of the business. The thief will later return, but may leave and return several more times before actually committing the theft. This arrive, leave and return action serves a useful purpose, as it allows the shoplifter to make the necessary preparations for stealing. For example, their first arrival is to ensure that the property they wish to steal is present and to check packaging or locate the desired sizes, styles, etc. Their second arrival may involve the removal of products from their packages or clothing items from their hangers or to stack several items together (including items from other parts of the store). At some point, the shoplifter may remove any tags bearing the store name from the property.

Shoplifting—Anything But Petty

Product Evaluation

Another sign of the characteristic demeanor of the shoplifter is a failure to look at products they are *holding*. They remove an item from the shelf, hold it in their hands, but never look at it. They may return the item to the shelf, select another, and do the same thing. Their eyes are never on the product because they are still focusing on personnel to see who is present and who is looking.

Certainly not all shoplifters operate in this fashion. Some are very bold and simply walk in, grab what they want, and walk out. This is especially true where the products taken are stacked or displayed close to an exit. Most shoplifters, however, are nervous and in trying to act normally, actually achieve the opposite effect. If an employee is trained to recognize these telltale signs, most shoplifters can be scared away and their crimes prevented.

The Means To Steal

When first discussing the profile of a shoplifter, it was noted that there is an exception to the "you can't identify them by appearance" rule. This exception is an indicator of potentiality termed *the means to steal*. When a female walks in to the business carrying an oversized purse or wearing what appears to be maternity clothing but doesn't look pregnant, or a male enters wearing a trench coat or other bulky clothing, there is reason for concern. Certainly, this alone is not sufficient cause to believe the individual is intent on theft, but special attention is warranted.

Tricks Of The Trade

Shoplifters are always trying to outsmart their adversaries by coming up with some novel way of concealing property. First, let it be well understood that it is amazing how much property a person can conceal on themselves without the property being readily visible to others. Remember, these individuals have little to do *but* plan. If they aren't stealing, they're *thinking* about stealing.

The "booster box" is a common theme among thieves. It is little more than a box or a bag with a false bottom. Here's the Scenario: You see the thief place the property in their bag, but upon post-arrest examination, you cannot find the property! Remove the false bottom and sure enough, there it is. There are as many variations of the "booster box" as there are personalities of shoplifters.

Shoplifting—Anything But Petty

Many crooks either save a bag bearing the store name from a prior legitimate purchase, obtain one from a trash can outside the store, or from behind a counter inside the store. Then they put their stolen items in the bag and attempt to walk out of the store as though they are carrying legitimately purchased items. Most of the time they'll keep the bag folded up in an interior pocket until they are in the store. In fact, it is a good idea to note in your report from where they retrieved the bag or even that the bag had creases suggesting that it had been folded. Never give a bag to a customer who has not made a purchase. By doing so, you may be giving him or her the means to steal. Get into a practice of bending the receipts over the top of the bags and then stapling them shut. Some businesses place a strip of brightly colored tape on the bag so that it is readily visible. This indicates that the items in the bag have been legitimately purchased. In a small store, however, this may not be of value.

Then of course there's the false pregnancy. In walks the female crook wearing what is obviously a maternity top. She could be pregnant, but she's not showing a great deal. By the time she leaves the store she'll have moved from "not showing" to "eight weeks overdue." The elastic waistband of her maternity pants and her smock will drape over whatever stolen property is hidden in there. Snap a photo of the crook as she looked upon arrest. Note the brand names of the clothing she is wearing. She may later say that she wasn't wearing maternity clothes, but rather "comfortable, loose-fitting garments." But with brand names like "Mother To Bee" or "New Mom," their classification is fairly obvious and the crook's intent well-established.

Female shoplifters will also use bundled babies as a means of transport, placing the stolen items inside the baby's blanket. Baby bags, strollers, carriers, and other similar items are certainly useful in shoplifting theft. Many of these shoplifters are also hoping that if they are caught, they can blame it on the child. Because of the child's young age, the crook is guessing that the child will not be prosecuted.

Whatever shoplifting women can do with maternity clothes or other loose-fitting clothes, men can do the same with overcoats, baggy pants, and boots. If there's an unfilled cavity, a thief will try to use it to hide stolen items. Be wary of customers who are overdressed for the weather, especially those customers wearing heavy jackets in warm weather and raincoats when there is no rain.

The fact that a crook prepared for the theft by acquiring a bag, wearing maternity clothing, etc., can raise the seriousness of the crime in some

Shoplifting—Anything But Petty

states. This action shows an intent to steal prior to entering the store. Check with the police agency in your jurisdiction to determine the statutory penalty. Always be aware that the property that gets out your door by way of shoplifting may also be brought back in "refund schemes."

Refund Schemes

Refund schemes are the best way to convert stolen property into cash money. The crooks steal items by way of shoplifting and then returns them (to the original store or another store in the same chain) for a refund. They will have a "sob story" about how the receipt was lost or was never given to them. Often, they're smart enough to hit your counter with property in a bag bearing your store's name. (Most likely, they found the bag in an outside trash can.) They'll contact your clerks when they are busy and do not have time to watch while the crook searches through a purse or billfold over and over again, looking for a receipt that they never had. Many stores prohibit any refunds without the original receipt. If they present a receipt, make sure it is for the same item being presented. Also, ask when the items were purchased and compare this information with the date on the receipt. If you "stir up the waters" sufficiently, the crook may get cold feet. Then, they grab the item, say "never mind" loudly, and stomp out of the store, acting like a disgruntled customer. Also, be careful of persons who say that they lost the receipt and wish only to exchange the item for a different size or color. They may be attempting to legitimize the transaction by asking if you could provide a duplicate receipt of some sort. Then, armed with the new receipt, they will attempt to return the obviously faulty or incorrectly sized product for a refund.

It may be helpful if your clerks are instructed to perform some sort of action on the item or the price tag of the item. For example, have them tear off a specific tag (or tear it in half) whenever the item is sold. Or have the clerks place their initials on the price tag. Items returned for refund that do not bear any initials were probably stolen, however there's always room for human error. Some major stores have devices that encode or decode strips on the tags of all merchandise. They can pretty much tell if an item presented for refund was ever sold. Again there is always the possibility of human error.

Till Taps

Another type of theft from businesses that is committed by patrons is the *till tap*. It's all very simple. The crooks generally work in pairs. One of them gets the clerk to open the till to pay for a small purchase, while the other does something to take the clerk's attention away from the register. They might drop something from a display onto the floor or simply ask for help. When the clerk is not looking, the crook at the register reaches across the counter and takes the larger bills.

Prevention of Shoplifting

While there are many steps that can be taken in an attempt to prevent shoplifting, the most effective measures require no additional money and only a little extra employee effort. Here are some steps you can take to prevent shoplifters from making your business a victim of their crimes.

Display of Products

Do not place items of high value or that are at high risk of thievery next to an exit or in areas where visibility is minimal. For example, do not display the video cassette recorders right next to the front door. It's too easy to just step in, grab a unit and run. When the business is closed, such an arrangement invites window smashing for the purpose of theft. The same is true with regard to "suitcases" of beer—leave them stacked by the front door and you're guaranteed to lose a few to shoplifters.

One business went to extreme lengths to prevent theft of VCRs. They improved the locks, put on iron grates, and even put pipes in front of the door to stop crooks from ramming the glass doors with their vehicles. Only when they moved the videocassette recorders from just inside the doors did they see a reduction in VCR theft.

Method of Display

Set up the display aisles so that you can see from one end to the other. For example, aisles set up parallel to the cashier's station do not allow clerks to see what is happening on the opposite side. Aisles set up in a perpendicular manner, however, allow the clerk to see from one end of the store to the other.

Cashier Location

Rather than setting the cashier's station deep into the store against an interior wall, choose a location that all patrons must pass in order to exit the store. This also increases the employee-consumer contact which is both good for business and good for preventing theft.

Shoplifting—Anything But Petty

Greet Everyone

Make a habit to greet everyone who walks through the door. Even if you are busy, let them know you'll help them if they need some assistance. If you're with another customer, politely excuse yourself and tell the new arrival that you'll be with them in a minute. This is absolutely your best weapon against shoplifters because they desire anonymity. When your eyes meet theirs, anonymity is lost.

Be Mobile

Try to move around within the store rather than being glued to the counter. Your unpredictability represents a nerve-racking barrier to the shoplifter.

Security Devices

It should be well understood that there is no security device made which can guarantee that shoplifting will be prevented. One can easily place thousands of dollars into price tag-activated magnetic alarms and video surveillance systems that will not be any better at preventing shoplifting than the five steps already discussed. There are, however, several products on the market that may be of value, without breaking the bank. Security systems are discussed in greater detail in chapters twelve and thirteen.

Wide-angle mirrors are an excellent means by which to see into areas of the business otherwise hidden from view. For the shoplifter, they represent the elimination of a perfect location to stand or crouch down while concealing property.

Residential burglars have been known to hit several houses on the same street, purposely bypassing homes having alarm boxes under the eaves. This would suggest that a homeowner, not able to afford an alarm system, might receive the same benefit by simply installing the bell box under an roof overhang. Likewise, business owners may be able to discourage shoplifting (and robbery) merely by presenting the *illusion* of having high-tech prevention equipment.

Several companies manufacture video cameras that outwardly appear to be quite real, but are merely a shell and have no internal workings whatsoever. Some even have an eye-catching red light on the top to add to their functional appearance. Models can be purchased that are motorized and pan from side to side. Nearly all have video cables that run from the camera to a wall plate and appear to enter the wall (but actually stop at the plate).

Some business owners fear that the presence of video cameras will cause legitimate customers to feel uneasy. These business owners might prefer to install plastic video camera housings that fit nicely into existing suspended-ceiling frames. These enclosures have a smoke-colored portal where it *appears* a video camera is situated. Another variation is the half-sphere that mounts onto the ceiling. Legitimate customers seldom notice them and if they do, generally do not know what they are. Thieves, however, are much more attuned to such devices and see them as a barrier to success.

One-way mirrors allow the person on the sales floor to see a reflective surface, while the employee on the opposite side is able to see through it as if it were clear glass. This is a great tool for watching employees suspected of stealing, but will work well for watching suspected shoplifters, as well.

Many devices on the market are designed primarily for robbery prevention, where a greater loss is generally sustained (and certainly greater risk to personal safety is present). Remote-activated alarms and self-sequencing cameras represent a significant investment and provide little value in the arena of shoplifting prevention. These devices are discussed in greater detail in the chapter on robbery.

When Shoplifting Is Observed

States differ on what action, if any, can legally be taken if a shopper is seen concealing store property beneath clothing, in a purse, etc. In some states, the act of concealment constitutes the completed crime of theft and detainment of the individual for police would be allowed.

There is an added benefit to detaining the individual at this time, as once he is out of the store he is more likely to flee—or fight (and he may have friends outside to help him). In other states, however, there is a specific requirement that the individual leave the store before the crime of theft is complete. The rationale here is that the individual could have paid for the item at any time prior to walking out the door and a theft would not have occurred. In some states, the law allows arrest of the shoplifter prior to leaving the store so long as he walked past the last cash register and failed to pay. The obvious question is "why would anyone hide property in their purse or under their clothing if they didn't intend to steal it?" *It is necessary to check with the law enforcement agency having jurisdiction where the store is located to determine the correct course of action.*

Another option not yet discussed is merely telling the shoplifter to put the property back and to leave the store. The subject may deny the theft, placing you into a position of having to confront the individual. This can create some difficulties if you happen to live in an area where the law requires the subject exit the store before a theft is complete. Or the individual may give the property back and apologize, stating that this was the "first time" he had ever stolen anything. In reality, most have successfully shoplifted many times before being caught—most likely, he will simply go to another store and steal.

The Decision to Arrest

A major premise of the American system of justice is that it is better that one hundred guilty men go free than one innocent man be wrongly convicted. This is a logical foundation upon which to base your decision to deprive a citizen of his freedom and accuse him of being a criminal. Opting not to make an arrest when you're not certain that a theft has taken place is probably the best decision you can make for your business. A $20.00 loss seems small in comparison to a two-million dollar civil lawsuit.

Making The Arrest

Some stores have their employees say: "I represent XYZ Company. Have you forgotten to pay for anything today?" This statement alone does not include any declaration of arrest, but rather is an attempt to encourage the thief to return the property and suggests that if the item is paid for, no further action will be taken. Some tell shoplifters: "I represent XYZ Company. Would you like to return our property or come back inside the store?" Again, it is suggested to the thief that if the property is returned, he will not have to go back inside where police will ultimately be called. The goal here is to get the property back and then take custody.

Most states require that the person making an arrest (often termed a citizen's arrest) advise the person of such. A common declaration is: "I represent XYZ Company and I am placing you under citizen's arrest for theft." It is important that the thief be told that you have the authority to make the arrest (as a representative of the store) and that you are, in fact, making the arrest. This may assist in dissuading later claims that he ran away or attacked you, not to avoid arrest, but because he thought he was being robbed by you.

Persons being arrested for theft may tell you that they must first go to their car to retrieve identification—if allowed to do so, they are unlikely to return. Should they inform you of facts that represent an emergency, such as a child left in a hot car, ask for a description of the vehicle and offer to send a store employee to rescue the child. *Again, when dealing with matters of law, it is absolutely necessary to check with law enforcement authorities in your area to determine what you are allowed to do.*

If a female is arrested by a male employee, it is highly advantageous to have a female employee remain in the room until law enforcement arrives. This will assist in countering any claims of sexual assault.

When Resistance Is Encountered

It is absolutely necessary that you determine what force may, by law, be utilized to effect a citizen's arrest in your area.

Always bear in mind when initiating an arrest, that for the perpetrator, there may be much more at stake than just a fine for theft. The thief may also be under the influence or in possession of an illegal drug which may result in additional charges once police are brought in on the matter. If on parole, a theft arrest may guarantee he will remain in jail until the case goes to trial and may ultimately lead him back to prison for violating his parole. *He may resist violently.*

Also very dangerous is the thief who panics at the prospect of arrest and is armed. Always be aware that anyone you contact in an arrest situation could be armed and that literally anything can be used as a weapon against you. For example, in one California case a store employee was advised by a shopper that an individual had secreted an $8.00 roast inside his pants. Employees watched the individual and once he exited the store without paying for the item, they moved in. He was taken into custody without incident (there were at least five employees in on the arrest) and taken to the rear office where he was detained for police. At one point during their questioning, a struggle took place and the accused pulled out a pistol and shot a teenage employee in the chest, killing him. He then fled from the office and shot an additional clerk prior to escaping.

Shoplifting—Anything But Petty

Though he was later arrested for theft, attempted murder and murder, the case was not as cut and dried as you might expect it would be. The defense claimed that store employees had no authority to make the arrest since they did not personally see the defendant secret the items onto his person (the actual witness left the store prior to the arrest being made and her identity was not known). Though this "defense" did not fly in California, it might have in another state. Additionally, the defense claimed that the store's use of five or more employees to make the arrest escalated the situation. It was alleged that the group of employees who made the arrest became a "gang of sorts," and as such were "de-individualized"—essentially, that they felt less responsibility for their personal actions and therefore felt free to use greater force to effect the arrest and detain the suspect. Citing evidence that the defendant was an alcoholic, the defense contended that his use of a gun was caused by paranoia and fear and that this paranoia and fear was caused by the "lynch mob" sent out of the store to bring the accused back inside. The accused was found guilty, but of the lesser offense of manslaughter, rather than second-degree murder. It is apparent that, to some degree, the jury bought the "gang" issue. This is why it is so very important to know the laws in your state and to think carefully before acting in such matters.

Do not let ego be the driving force behind a wrestling match with a thief. It is possible to win and still lose, just as it is possible to lose and still win. Every employer should have a written policy regarding shoplifters that resist and fail to cooperate. When deciding just what that policy should be, the employer should first take a good look around the store and ask if there is anything on any shelf, anywhere in the store, that is worth being seriously injured or dying for. Failure to develop such a policy and enforce it consistently may result in you being sued by your own employee for injuries suffered by that employee in an attempt to make an arrest. After all, what training have you provided them in making arrests? Overcoming resistance?

Checking For Offensive Weapons

Some states allow private citizens making citizen's arrests the authorization of conducting "pat-down-type searches" to insure the person being arrested is not concealing a weapon that might be used against them. An inquiry should be made to the local law enforcement agency or the county or city prosecutor's offices to determine whether or not such a search is allowed in your state.

If a pat-down search is allowed by law, it is limited to the exterior of the clothing. Items believed to be offensive weapons may be seized and

Shoplifting—Anything But Petty

turned over to the law enforcement upon their arrival. It is best that a female be checked by a female employee to reduce the likelihood of sexual battery charges. Likewise, male shoplifters should be checked by male employees whenever possible.

With regard to concealed store property, time is on your side. For example, if the property has been placed into a purse or concealed upon the person, it is better to await the arrival of law enforcement. In this way, the law enforcement officer can identify from where the property is recovered, thus corroborating your observations. If the property was secreted to a purse, maintain custody of the purse until police arrive. The same would be true of any bags, boxes, or other containers which were used to facilitate the theft.

Identifying The Offender

Recognize that in most cases, the first name a shoplifter gives you will most likely not be their true name. And, while you should go ahead and write down the name they give you, be aware that you may have to either complete new documents or place the correct name on the same form. Do not obliterate, erase, or otherwise remove the false name given—it is valuable evidence, as lying shows consciousness of guilt. Your records can then reflect not only the true name of the offender, but also the alias or aliases used. It serves another purpose, as well: most crooks don't have a name prepared to give you. When you ask, they quickly come up with something. Often this is a combination of their true first name and a false last name. They may reverse their first and last names. They may give the name of a friend—which is a mistake. (One subject gave the name of a friend and was promptly arrested—he didn't know his friend had outstanding warrants!). Whatever name they give, they may not be able to remember it the second time they are asked. This is why it is important that you always write the information down, but keep it so that the crook cannot see it. The same holds true for the crook's date of birth. Often they just make up a number, or give a brother's date of birth. Write down whatever they tell you. Later, when asked their name and date of birth again, they will not be able to recall the information they have given you and their lie will be obvious.

A good police officer will take the information sheet upon which the store representative has written the name and date of birth given by the crook. The officer will then look at the form and verbally confirm with the crook that the name and date of birth he or she has given to the store representative is correct. This is done because in most states it is illegal to give a false name to a police officer, but not to a private citizen. In a minute or two, the officer will ask the crook to give his name and date of birth again. Usually, it is different than what was given the first time—too much time had elapsed and the crook couldn't remember the information he had given the first time!

Keep a close watch on detained shoplifters. They'll often try to stash their wallets (which contain their true identification) somewhere nearby. The same is true of contraband and weapons. In many cases, the crook will have warrants of arrest on file and does not wish to be identified for that reason. In other cases, the crook is simply hoping he can give a false name and then be released on a citation (or other promise) to appear in court. He will never appear and a warrant will be issued, but not for the true offender.

In most states private security officers and/or persons making private citizen's arrests are allowed to go into bags, purses, etc., to retrieve their property. The scope of this action may be limited by the law in your state. For example, the law may allow you to retrieve property taken from the store, but nothing else. Another state may allow you to retrieve property and identification. Make sure you understand what the law in your state allows you to do. If you are unsure of what the law is, it is best to wait until a police officer arrives. Let the officer search the suspect. Of course, if your state says you may retrieve property taken from the store but search no further, just make sure your report reflects that what you saw came as a result of retrieving the store's property and not a part of some continued "search." Recognize that any actions of this sort on your part provide an opportunity for crooks to claim that *you* ripped *them* off. They claim the police rip them off all the time.

The Importance of Documentation

After the event and as soon as possible, begin making notes of what has occurred. Keep in mind that it may be months (or even years) before you will be called to court to provide testimony at trial. Delay tactics are almost always used by the attorney representing the accused, in hopes that witnesses will move, die, or forget. As well, accurate documentation of your observations and lawful actions may prevent the filing of—or reduce the harm of—a civil lawsuit.

There is a tendency to believe that reports written by private persons making arrests should look and sound like reports written by police officers. However, it is quite to the contrary. In fact, law enforcement is reassessing the way they write reports, moving away from vague terminology and police jargon to simpler, more understandable stories of what happened.

Reports written by store personnel should minimally include an explanation of what first brought the perpetrator to their attention. If it was reported by another employee or a patron, their names and statements must be included in your report. If it was something the perpetrator did, such as selecting items off of the shelf but looking around at store personnel rather than at the item, the actions should be documented. Again, *be specific*—don't say the person "acted like he was going to shoplift." Rather, indicate the specific action, describing what you saw. You can also relate the person's actions to your experience with other shoplifters and may use this as justification for continued observation. Anything and everything that the thief did (or didn't do) that leads you to logically believe he had committed a crime should be fully documented.

Describe the actions of the perpetrator, from the moment you first became aware of him until you formally detained him for theft. Include anything said once contact is made. Documentation that the individual claimed, "I had to steal the stuff, my kids don't got no clothes" may prevent a later claim that the items were properly purchased.

If there was another employee present at the time of the arrest or standing by with you awaiting the arrival of law enforcement, be sure to include that individual's name in the report. A second witness can assist greatly in refuting claims of sexual assault or abuse. Bringing up the name later, after such claims have been made, may become a credibility issue.

Maintain all notes and reports for a minimum of one year, even if the thief pleads guilty. In most states, one year after the event is the time limit for filing civil actions.

EXAMPLE:

September 5, 1993, 9:20 P.M. I saw the white male, 35 or so, wearing a red sweater and blue jeans in aisle 10 (electronics). I saw him take a roll of black electrical tape off of the display and hold it in his hands. He did not look at the item, but instead, began looking left and right and up and then back over his shoulder. He then put the roll of black tape into his right front pants pocket. I followed the man out of the store and once outside, told him I was with store security and asked him if he had failed to pay for anything today. He said,

Shoplifting—Anything But Petty

"Okay, you got me. Can't we work this out?" I told the man he was under citizen's arrest for shoplifting and asked him to accompany me inside the store. He went with me into the office and the police were called. When the police officer (Officer Jones) arrived, the man removed the black tape roll from his right front pants pocket and placed it on the table. I examined it and saw that it had our store sticker on it. I wrote my initials on the tape package and the police case number. The item is being kept for court in secure storage. The man told the officer (Officer Jones) that his name was John Smith.

Certainly, this would not be the totality of the clerk's report. However, this should give you some indication of the type of documentation that is necessary.

Maintenance of Evidence

The property taken and any bags, boxes, etc., used to conceal the property are the evidence in a shoplifting case. In some jurisdictions, law enforcement will seize the items and place them into their evidence system until trial. Increasingly, law enforcement is relying upon merchants to maintain the evidence themselves, as their systems are heavily overburdened.

To prove that the item produced in a June courtroom trial was the same stolen by the thief in January, the arresting citizen's initials should be written somewhere on the package/item. The marking can be small and in an inconspicuous location—so long as it can be pointed out in court as proof of the authenticity of the item. Placing the item back on the shelf to be sold and selecting a similar item to take to court when the trial comes is not only unethical, but a very dangerous game to play. This action could result in a mistrial (where all proceedings against the accused are dropped) or an outright acquittal. The next step might be criminal charges against you for falsification of evidence and/or perjury (lying under oath). It may also generate a civil lawsuit for false arrest and a number of other civil rights violations.

In addition to initials, you may wish to include either your store event number or the law enforcement case number—or both. Once the case is completed, you may return the item to the shelf.

Testifying In Court

In the vast majority of cases, those persons arrested for shoplifting crimes plead guilty. There are always those who will test the issue, generally because they cannot afford not to. Persons whose standing in the community will be tarnished if they admit to such a crime and individuals whose criminal histories guarantee considerable jail time are the most likely candidates for a not guilty plea.

Shoplifting—Anything But Petty

Upon the receipt of a summons (also called a subpoena) to appear in court, attempt to contact the prosecuting attorney who will be handling the case. Take a few minutes with the attorney to discuss the case and what will be expected of you.

Arrive early at court, even though you may wait several hours for your case to be called. Even judges have bad days and have little patience for witnesses who are tardy. As you wait for your case to be called, do not engage others in conversation about the case. If someone attempts to talk to you about the case, politely reply that it would not be "proper" to discuss the case prior to your testimony. The defense attorney or one of her investigators may be searching for information they can use to damage the case or to destroy your credibility. They will have ample time to ask questions of you while you are on the stand (but there are fewer rules in the courtroom lobby). Naturally, discussing the case with the prosecuting attorney is acceptable, so long as it is done out of the earshot of the adversary.

Just prior to testifying, you will be administered an oath, wherein you promise to tell the truth in the matter now before the court. When asked if you agree to do so the correct reply is: "I do." Once taking the witness stand, you will be asked to identify yourself and your occupation. In a series of questions, the prosecutor will lead you through the theft, from first observation to arrest. Answer the questions fully. Then, the defense attorney gets a chance to try to discredit your memory, your eyesight, your report—and ultimately, your testimony.

No matter how rude the defense attorney may become, maintain your professional demeanor. She may be trying to make you angry so as to damage your credibility (your believability) with the judge and/or jury. Always pause for a moment before answering any question posed to you by the defense attorney. This gives the prosecutor a chance to evaluate the question and register an objection if it fails to meet the requirements of the law. When answering questions for the defense attorney, do not elaborate. Simply answer the specific question (with a "yes" or "no" if possible) and do not tack on explanations. The defense attorney may be fishing for new information to use against you.

It is perfectly acceptable to not look at the attorney asking the question when answering. Looking at the defense attorney and her well-practiced, intimidating body language can create nervousness. Most successful police officers look at the jury when answering defense attorney questions. By picking out a few friendly faces and talking directly to them, a relationship can be established. Plus, the intimidation factor is essentially nullified.

And while it may sound as if the court trial is some sort of a game, rest assured, it is not. Somebody's freedom is at stake, not to mention their reputation within the community. Answer all questions honestly. If that "honest" answer means you have to say, "I don't know" or "I can't recall—so be it. And don't let the defense attorney pressure you into giving an answer. For example, if the defense attorney asks you how many feet you were away from the defendant when you saw him hide the property on his person and you don't know, simply say, *"I don't know."* The attorney will now hit you with a series of rapid fire questions, such as: "Twenty feet? Thirty feet?" etc. Your only response is the first one you gave and that is: "I don't know." You may be able to provide an answer by using the courtroom as a tool. For instance, you may be able to say, "I don't know how many feet it was, but it was about the distance from your table to where I'm sitting." But, again, never let a defense attorney pressure you into answering a question when you don't know the answer or didn't see or hear that what she is questioning you about.

Your job as a witness is to provide testimony about your personal observations, knowledge and involvement in the case—not to try and win the case single-handedly by making the perpetrator and his attorney look ignorant. In fact, smart remarks and witty quips have no place in a court of law. If you forget, the judge will most likely remind you.

In Conclusion

The best shoplifting prevention tools are your employees and customer service—not electronic gadgetry. Shoplifters almost always give themselves away through their demeanor. It is of extreme importance that every business owner determine just what actions they can legally take in their state if shoplifting is observed. Arrests for shoplifting should never be made in a case where you are unsure if a theft has taken place. When arrests are made, you can both strengthen the case and protect yourself by insuring the event is thoroughly documented. Shoplifters may become violent and resist—your health and well-being must always take priority over an arrest. *The best arrest, remember, is the one you don't have to make, because the crime was prevented.*

Chapter Four

Armed Robbery

Some businesses are more susceptible to armed robbery than others. Businesses that stay open later in the evening or for 24 hours are robbed at a greater frequency than those that shut down at five in the evening. Darkness seems to give the armed robber a better chance of not being seen, a better chance of escape, and a lesser likelihood of witnesses to the crime. Realistically, all businesses are potential targets, regardless of the time of the day, and therefore, all must pre-plan for this eventuality.

Armed robbery is the most life-threatening crime that business owners face. Not only are the suspects armed, but they are also more bold and usually more desperate than other types of criminals. The weapons they carry may be knives, handguns, shotguns or nothing more than a toothbrush in a pocket, simulating a weapon. Whatever the weapon is, the best thing to do is to cooperate. The bandit will want to get in and out of your business as quickly as is possible. Do exactly as they tell you and don't argue with them. Many times these people take drugs just before a robbery to build up their courage. Others have major ego problems, which are eliminated by the gun they hold in their hand—and they may opt to use it in order to meet any challenge sent their way. The guys (99% of them) are time bombs just waiting to go off—don't give them a reason.

Armed Robbery

Most businesses have a "drop policy," where large bills are hidden or placed into the safe. This is a good idea, but most people, including crooks, are aware of this. It is not at all unusual for a crook to have the register emptied, then demand "the rest of the money." This of course, is reference to the "drop money." This is where they can really get nasty if they don't get what they want. Telling them that there isn't any more money or that you don't have access to it (for whatever reason) may set them off. It is a good idea to keep a small amount of money in an envelope or metal box close to the cash register. Place a twenty dollar bill on each side of twenty ones, and then rap them up with a rubber band. If the suspect demands "the rest of the money," you can always give them this. Hopefully, this will prevent them from becoming violent and also protect your real "drop money."

Many business have electronic safes that will allow them to obtain a set amount of money at specified intervals. The drop money is immediately placed down a slot and is inaccessible. When additional money is needed for the register, the clerk may have to wait for the next "delivery time" when the safe will give them additional funds.

Whatever you do, make sure you tell the suspect what you're doing. For example, you might say, "Okay, the rest of the money is under the counter, I'll get it for you." Just reaching underneath the counter may cause the crook to believe that you're reaching for a weapon or a robbery alarm button. Nobody wants to die a victim of a violent crime, but certainly it would be an injustice upon an injustice to be shot or stabbed simply because you were doing what the robber was demanding of you. Try to stay as calm as possible and do not make any sudden movements that the robber could interpret as being a threat.

As soon as the suspect leaves, call the police. Don't call your spouse or anyone else before calling the police. (Employees should be instructed NOT to call you before calling the police—this is quite common!)

If you have a robbery alarm and you have not yet hit it, now is probably NOT a good time. An alarm signal tells the police that the alarm works—that's all. It does not verify in any way that an actual robbery is occurring (although responding officers will treat it as real). Alarm signals cannot provide the suspect description, vehicle description, or any direction of travel—*you, in some cases, can*. Determine from your alarm company whether or not your alarm goes to an answering service, which then calls the police, or whether it goes directly to the police. Additionally, find out if your alarm has its own dedicated line, or if it is linked through your phone line. If it is linked through your phone line, once you have activated the alarm you will not be able to call the police because the line will be tied up! When suspects get what they came for, they generally leave in a hurry (despite their threats of "I'll be watching you—if you try to call the cops, I'll shoot!"). Using the phone is probably safe after you hear them hit the door. If, for example, you were ordered to lie on the floor and are not sure if the suspects have left, it may be inadvisable to move to the phone. If there is a robbery alarm button that can be reached without standing up, this would be the wiser choice.

Once you make contact with the police, time is critical. The police will ask for specific information—try to be calm and answer their questions as honestly as you can. Some questions may seem somewhat unnecessary (because the officer can just ask them when he or she gets there), but there is a reason for each and every inquiry. In most cases, a dispatcher will send a police unit within seconds of answering your call. They will keep you on the line, however, in order to get additional information to relay to the responding units, and also to try and help you survive the traumatic ordeal. Remember that dispatchers handle emergencies routinely and that their voice will sound calm—almost like what you're reporting isn't a big deal. The truth is that any excitement communicated in their voice would only serve to further excite you, which can reduce the quality of communication achieved.

Depending upon how far away the units are, the response time to your business could be under a minute or as long as twenty minutes. Inside an incorporated city, the response times will almost always be much less than if you reside in a rural area. City police have traffic and traffic lights to deal with and county police may have twenty miles to cover just to get to you.

While the dispatcher has you on the phone, you may hear some typing in the background. The dispatcher may periodically move her mouth away from the phone and answer requests or provide additional information to the units responding. Do not be concerned about how long the dispatcher is keeping you on the phone. Although some people believe that units are not sent until the phone call is concluded, this is not true when dealing with in-progress/just-occurred crimes.

Helping Police Locate The Suspect

Here are some of the questions you may be asked:

1. What is the name of the business?
2. What is the address?

Armed Robbery

3. What is the nearest cross-street?
4. Is anyone injured?
5. What is the description of the suspect or suspects?
6. What kind of clothing were they wearing?
7. Were the suspects armed? If so, with what? If a gun, were shots fired?
8. In what direction did the suspect flee?
9. Was there a vehicle seen or heard? If so, what is its description?
10. What property was taken (general description, i.e., money, guns, etc.)?

With regard to the description of the suspect, expect to be asked:

1. Race?
2. Sex?
3. Approximate age?
4. Height?
5. Weight?
6. Build (muscular, fat)?
7. Color of hair?
8. Length of hair?
9. Facial hair?
10. Glasses?
11. Scars?
12. Marks? Tattoos?
13. Hat?
14. Jacket?
15. Shirt (markings?)
16. Pants?
17. Shoes?
18. Odors-Missing teeth-Jewelry?

With regard to the vehicle, expect to be asked:

1. Year?
2. Make?
3. Model?
4. Body style?
5. Color(s)?
6. Damage?
7. Add-ons?
8. Loud mufflers?
9. Lights out?
10. Occupants?
11. Direction of travel?

Armed Robbery

With regard to the weapon used, expect to be asked:

1. Weapon seen?
2. Weapon simulated/not shown?
3. Pistol/Rifle/Shotgun?
4. Revolver/Semi-Auto/Pump?
5. Finish?
6. Barrel length/sawed off?
7. Caliber (if known)?
8. Weapon fired?
9. Where did the suspect carry the gun? Holster? Waistband? Boot?
10. Hand used to hold (left or right)?
11. Gun taken or left behind?

Naturally, these are only a sampling of the questions you might be asked. Some questions will be asked by the dispatcher at the time of the original call, while others will wait until after the arrival of the officer and the stabilization of the scene.

Fear does terrible things to our perception of time and our ability to see and recall. It might be wise to provide your employees with a checklist of identifying characteristics, so that they can be better witnesses. They also can upgrade their observational skills by testing themselves. Police officers do this by looking at someone and then looking away. Then, they try to describe the person's appearance and clothing. When they think they have it right (or are stumped), they look back and check to see how they did. Your employees can do this with customers—look at them when they first walk through the door and then try to describe them in their minds. When the customer makes it to the counter with their purchase, they can then check to see how close they were.

One good idea is to have a mark on the door frame (inside) at the five foot level, five and one half foot level, and six foot level. This gives the clerk a frame of reference when attempting to determine the height of the suspect. When he heads out the door, the clerk compares his height to the markings and arrives at an approximation.

Another good idea is to keep a bottle of glass and tile cleaner next to or under the cash register and stress to your employees the importance of cleaning the counter often. Not only does this keep the counter looking good (and that's good for business), but it also reduces the number of fingerprints that will be on the counter top. Treat the glass doors in the same way. There's nothing a cop likes to hear better than, "I just cleaned the counter before he came in—any prints should be his."

Armed Robbery

Try to pay attention to what the suspect touches. Often, the suspect will "browse" in the store while he waits for customers to leave and ensures that there isn't another clerk in the back someplace. In many cases, the crook will bring a product up to the counter and use the purchase as a ruse. Once he gets close enough, he takes one more look around and then initiates the robbery. In many instances, while awaiting the arrival of the police, clerks have been known to replace the item back to the shelf and never tell the officer anything about it. Undoubtedly, the shock of the robbery plays a major role. In one case, however, police retrieved an identifiable print off of a bag of dog food. The suspect, when caught, had a great deal of difficulty in explaining how his fingerprint came to be on an item, used in a robbery, at a store he testified he'd never entered in his life!

At the time, if it is safe to do so, watch how the suspect leaves. Did he put his hand on the cross bar on the door or the door frame? Or, did he do what many do and place his hand flat on the glass as he busted out the door in a hurry? In one case, police found an entire hand print—the entire palm and all five fingerprints—on a glass door that had just been cleaned by the clerk. This was not so much great police work as it was good practices and good observation on the part of the clerk.

Once the suspect is gone, lock the doors. You are effectively securing the crime scene and all of the evidence that might be there. Do not touch any of the surfaces that you know or suspect the crook touched, nor allow anyone else to do so. If there were customers in the store at the time of the robbery, ask them to remain. If they cannot wait, for some reason, try to get their name, date of birth, address and phone for the police. If they refuse to identify themselves, try to get the license plate of any car in which they might depart. The customer may not wish to get involved in a lengthy court trial. Or the customer may be the crook's partner, sent in to "scout" out the place and be there as a back up in case some clerk or a customer decided to be some sort of a John Wayne. Sometimes, they'll even stay behind and give information that differs greatly from that given by the other witnesses. Of course, this is merely an attempt to throw the police off track and cause them to consider the suspect taller, shorter, of a different race, or driving a different make or color of car. If you are later called to view a photo line-up which includes the suspect and are able to pick out the bad guy, this "witness" will refute it, either picking out another one or saying, "he's not in this group of pictures." Most robbers are not this sophisticated, thankfully. Generally, this "shopping partner" is there strictly to check out the place and, if he sees something he doesn't like, he merely exits the store and warns the partner who is waiting outside.

Armed Robbery

Helping to Prevent Armed Robbery

It is important that the business be well lit and your cash registers close to the entrance. Lots of glass is always best, as it allows cops to see in and doesn't allow crooks to feel comfortable. If the crook thinks he can be easily seen by passing motorists, he'll pick another spot. Trouble is, most businesses are out to sell a product and the best way to do that is to advertise. Windows get covered with sale posters advertising the cost of beer or toilet paper; tall displays and magazine racks also may block the view in or out. There must be some compromise between your safety and sales.

If you have a video camera aimed at the door or the register area (or both), make sure that your displays are not blocking the camera's view. Lighting is especially important. Too many times, videos show that a robbery certainly took place, yet provide little in the way of identification because of obstructions or poor lighting. Increase the lighting, if necessary, and move displays. You may also be able to "open up" the camera's aperture (like the lens of an eye) to brighten the picture. Best way to tell? Test the system. Double check the angle—does it show the subject sufficiently to identify him by facial features? Is the camera too far away or the lens too wide-angled to provide positive identification? If the camera can't identify the crook, the only value you can attribute to it is that it will show that a robbery did take place and provide a very general description of the suspect and his movements.

Scanning radio receivers are an interesting and sometimes highly valuable tool in robbery prevention, but are not without some risk. These radios monitor a multitude of police frequencies but, of course, do not allow you to transmit on any channel. Situated in the back room and cranked up loud enough, a radio scanner may cause a would-be robber to think that some cop is in the back room sucking down a cup of coffee and a few free doughnuts. He'll be thinking—did we miss a patrol car outside? In any case, the hope is that the robbery will not take place. One negative aspect of this idea is when your business has a silent alarm and you activate it while the suspect is still in the store. It is entirely possible (unless you tune the scanner to a single channel of an agency not serving your immediate area) that the crook might hear the robbery alarm being dispatched. Once he realizes that you've set off an alarm, he may become angered and utilize the weapon in his possession to vent his anger.

Armed Robbery

If you have a robbery alarm, use it only for robberies. It should never be used to report disturbances, car collisions, shoplifting crimes or drunk drivers. Misuse of the alarm could result in fines and/or revocation of your alarm permit. If anyone is injured as the result of your improper use of the alarm, civil and criminal penalties could result as well. Remember, if you activate a robbery alarm, the police will be dispatched on an armed robbery. They are going to "prone out" any of your customers who are leaving the store upon their arrival and hold them at gun point. Where there is confusion, there is always the opportunity for Murphy to visit (referring to Murphy and his "laws"). How awful it would be if a customer were shot and killed by police because he didn't understand why he was being commanded to lay down on the asphalt parking lot and maybe reached for his wallet to grab his identification. It was a scenario that you set up by misusing the alarm. You will have to face the circumstances that result.

When police respond to a robbery in-progress, they will attempt to confront the suspect after he leaves the store. This generally keeps the crook out in the open and reduces the likelihood of a hostage incident. They will park their cars so they cannot be seen from the business. Once the suspect is confronted, *anything* can happen. If you hear gunfire, hit the floor and stay down. If you can hear the police commanding the suspect to stop, try to get to a back room and lock yourself and any customers inside. This will hopefully prevent you or others from becoming a hostage.

Hostage Situations

You actually have a better chance of winning the lottery than you do of becoming a hostage. We all know how to try to win the lottery, but few of us know how to be a hostage. There are actually steps that you can take to reduce the likelihood of being injured or killed during a hostage situation. If you are taken hostage, your chances of survival are very good. Police SWAT teams have hostage situations down to a science, where their activities are more psychological than they are action-oriented. They call it "the three T's," which stands for "time, talk, and tear gas." They will attempt to establish a dialogue with the crook and entice him to recognize the futility of his situation and that he is at risk of increasing the charges against him.

This is a highly traumatic period for both you and the suspect. During traumatic times, it is quite easy to "bond" with those around you, even strangers—and even robbers. It is a scene that has been played many times. In fact, this has happened so many times that they have given it a name: "The Stockholm Syndrome." The longer you are a hostage, the less likely you are to be injured or killed. Support the crook. The crook starts to like you and, to some degree, it is easy to feel sorry for the suspect (considering what lifestyle or needs caused him to resort to robbery). There have been incidents where hostages have had the opportunity to escape or even shoot and kill the suspect, but have chosen not to do so.

The police will use time to their advantage. Rarely will they immediately storm the business—that nearly guarantees that someone will be injured or killed. There may come a time when such will be necessary. When that time comes, SWAT may use diversionary tactics, such as flash-bang/concussion grenades to assist them in making a clean entry. At such a time, the object is to take out the suspect and avoid injuring the hostages. Time, however, is generally on the side of the police. A trained police negotiator can usually convince the robber (or the suicidal subject, whatever the case may be) to give himself up. The relationship that you build with the suspect can be of great assistance. Rather than condemning the crook, support him. Be critical of the police. Talk about how they love situations like this so they can use their high-power guns and "get a kill." Tell the crook, "Don't let them kill you—it's what they want." If you think you are irritating the suspect, however, keep your mouth shut.

If the suspect tries moving you from the business, a police sniper may try to kill him with a head shot. Experience has proven that allowing the suspect to remove the hostage is far more dangerous to the life of the hostage than attempting to take out the suspect with a head shot. If you are being used as a shield, make yourself as small as possible. If you hear a shot, drop to the ground and don't move until someone moves you or you are instructed by the police to move.

Keeping A Gun In Your Business

You may want to consider keeping a gun in the store for defensive purposes. The Constitution of the United States originally granted you this freedom. Laws vary from state to state, as do interpretations of the Constitution. With the placement of a gun in your place of business comes great responsibility. You must make sure that you are fluent in the laws of the state in which you are conducting business and how they relate to the maintenance of firearms in businesses and the use of deadly force.

Armed Robbery

You must ensure that you are properly trained in the use of the weapon—certified, if possible. And if your employees are allowed access to the weapon, you must see to it that they have been afforded the same training and understand the laws of the use of deadly force. They are your employees. You have provided a firearm for their use, and you are responsible for their actions. If you elect to keep a firearm in the store, carefully consider the capabilities of the weapon you have chosen. Your aim is to stop the perpetrator, not kill. However, you very well may kill in the process. Be certain that you are morally and legally justified to kill before you ever bring that weapon out. Don't plan on "winging" the crook or shooting the gun out of his hand. Your respiration rate and fear prevents such "trick shooting" (there's no fear on television—that's the only place this kind of shooting is ever done). Ask yourself, "Is there anything in the register worth dying for? Is there anything there worth killing someone over?" One ends your life and the other changes your life—forever. Few cops are ever the same after having to take the life of another human being.

If you're going to keep a gun at your business, you may want to have it hidden in a back room. If a robbery suspect is going to kill, they will frequently take the victim into a back room. If you are being herded there by a gun-wielding suspect, usually the matter at hand has just become much more serious. Have a metal box with a few bucks in it hidden close to the gun. Offer to get it for him. Have a gun hidden nearby. If he allows you to get the money, you might have a chance to go for the gun. Try to grab the box with one hand and the gun with the other and turn so that the first thing the suspect sees is the box. With a little luck, it will also be the last thing he sees. If the suspect directs you to move away and says he will get the money from the box himself, be sure that he won't also be able to easily see the gun.

There is a warning in order. If you are a well-practiced, highly-accurate marksman who is quite familiar with short range combat shooting, you may stand a chance of a success. You may also die. The ability to shoot accurately under stressful conditions is a rather unique ability. If you don't have these necessary traits, you may cause your own death by going for the planted gun. Certainly if you believed strongly that the crook was moving you to the back room specifically for the purpose of killing you, the decision would be much easier to make and the odds less important.

If you have a gun or guns in your business, have them hidden and have a plan for getting to them. Most importantly, know the laws in your state for their use and become proficient with the operation of the weapon.

Armed Robbery

Robberies Outside Of The Business

Robberies do not only occur at the business, but also when the owner or manager goes to the bank to make deposits. A money or bank bag is the worst thing in the world to carry your money in. You might as well carry a large neon sign that says, **"I'm carrying money—rob me!"** Don't advertise that you're going to the bank. Carry your deposits in a lunch box or paper sack. Before you take any money out of the business, go outside and look around. If you see someone suspicious (such as someone just hanging around without any apparent business, two guys parked in a car with the engine running, etc.), call the police. If all appears clear, then take the money to your car. If possible, get someone to follow you to your car. Don't walk together, stay twenty to thirty feet apart. This will cause a would be robber to think twice. If you're together, he'll just rob both of you. Better yet, (if possible) one person should drive the car as close as possible to the business; the second person can walk right out of the business and hand the cash to the driver. While in the car, keep the windows up and the doors locked. This will keep someone from jumping into the car while you are stopped at a stop sign or a red light.

It is a good idea to drive a different route each time you go to the bank. It is also wise to vary the time of day you make your bank deposits. If you are going to the bank and believe you are being followed, make one complete trip around the block—you can almost always spot a "tail" by using this trick. If you see a car behind you making the same four turns you have just made, there is a good chance that you are being followed. Know where the nearest police station is and drive directly there. There is almost no chance that a would-be robber would follow you there. If you go to the bank while it is open, use the drive-up window (many will not accept rolled coins—save these for the next day). This prevents someone from running up and grabbing your deposits from you while you are walking up to the bank.

If you make night deposits, a higher level of care should be exercised. When you arrive at the bank, drive around it a couple of times. If you see someone suspicious around the bank, make your deposit the following day. If you're taking a large amount of money to the bank, have someone take a second car. The other person should arrive just before you do. That person should get out of their car and stand next to it while you walk up and make your deposit. A would-be robber may believe the second person is acting as a guard who may be armed.

Armed Robbery

The key? Be aware of what's happening around you. Use common sense and trust your intuition—if something doesn't seem right, it probably isn't. Don't be afraid to call police—if people can call constantly about barking dogs or a neighbor's tree that drops leaves on their lawn, certainly you can call if you believe you're being watched and/or set up for a robbery.

Chapter Five

Commercial Burglary

Commercial burglaries represent about one-third of all burglaries committed in the United States. The burglar is generally a white male (70% of the time), under 25 years of age with a prior arrest record (80% of the time). In most cases, the burglar has no desire to have contact with anyone—they'd much prefer to get in, get what they want, and get out. This is why most residential burglaries occur during the daytime (when people are working) and commercial burglaries occur at night (when businesses are closed). All in all, the crime of burglary represents the crime with the greatest amount of success, but with the least amount of risk. Even if caught, burglars are not likely to receive heavy penalties, such as in robbery, where a weapon is used.

Commercial burglaries are generally not "spur of the moment" decisions. Rather, they are generally planned out to some degree. The burglar that grabs an opportunity to commit burglary without any prior planning is generally an amateur.

The goal of the crime of burglary is most often one of economics. If a drug addict has a one-hundred dollar per-day habit, he will need to steal somewhere in the neighborhood of four hundred dollars worth of property—a 25% exchange rate.

Commercial Burglary

How Burglars Enter Your Business

Businesses are at great risk for the crime of burglary. Many of the same conditions present that allow you to stay in business are the same conditions that leave you vulnerable. One example is that of large glass windows—the intent of their presence is to attract customer attention and hopefully sales, but they are open doors to late night burglars.

Though in some cases, burglars have been known to hide in the store and commit their crimes after the proprietors leave. The two most common methods are: "window smash and grab" and rooftop-entry.

Business owners should consider crime prevention when deciding what type of landscaping to use around their store. The store that decides on using large rocks in their landscape usually takes them out (after the fourth burglary where these same rocks were used to smash windows). Bars on the windows may prevent this type of entry, but they do little to enhance the aesthetics of your business. Don't provide the burglar with the means to victimize you—that's too easy. Additionally, tall bushes around doors and windows provide the burglar with a working area out of view.

One common method of gaining entry is to take a stolen vehicle and back into the doors of a business at a speed sufficient to cause massive glass breakage and, if present, render metal bars and grillwork worthless. Many businesses have had to place six-inch diameter pipes, filled with concrete and buried deeply into the ground, several feet out from the doors.

Things You Can Do To Discourage Burglars

Establish a policy not to place high-value, commonly stolen items at or near entrances, exits, or windows. A burglar can smash a window, grab a $400 video cassette recorder and be gone in only a matter of seconds. Shoplifters practice a "grab and run" that is usually successful. Set up your store so that these popular items are well inside, making their theft more difficult and reducing the likelihood you being chosen as a victim.

Leave your cash register open, even if it may still contain a small amount of coinage. This shows the thief that there is little or no money present there. Too many business owners have had to replace registers that were damaged by prying the drawer open. The cost of a new register far exceeds any loss of a few coins.

How easy is it to gain access to the roof of your business? See that metal doors are locked across any affixed ladders. In many cases, burglars have been known to climb to the roof using pipes that run along alley walls. This is nothing that a can or two of white grease won't prevent (it works!). Keep in mind, however, that if your building shares a common wall with others, your efforts may be in vain, unless they join your prevention effort.

On the roof, inspect any old air conditioning vent openings. Sheet metal is easily pried open, allowing access to the interior. In fact, some crooks have been known to knock air conditioning units off of their bases and have gained entry in this way. If you have a roof access door (from the interior of the store), be sure that it is made of quality material and is secured, preferably from the inside. Any alarm system should include contacts at this point, or some sort of a light beam or other device that activates if entry is made through this opening.

Insure that the serial numbers of the items you have on display are logged somewhere, so that in the event of a burglary, you can provide them to law enforcement. Many store owners write the serial numbers of display items on their respective boxes, noting on the box that the item is on display. The boxes are then kept in a special area of the store. Without a serial number, it is difficult to recover property and make arrests for theft and/or possession of stolen property.

Always take a complete walk through your place of business prior to locking up. One easy way for crooks to get in is to enter legitimately and then hide in the store. If you have an alarm system that uses motion detectors, the alarm is often set off shortly after closing. Recognize that this could mean someone is hiding inside the store. Be careful about telling the police that they need not respond.

Chapter Six

Bad Checks

If you run a business, accepting checks is a fact of life. It is also a fact that, from time to time, you'll get stuck with a bad one. Most people believe that passing a bad check is a criminal offense and that all that needs to be done is call the police. In some cases, this is true. In many cases, however, the police simply will not be able to help you.

In most states, *criminal intent* is required to commit a crime. If someone writes you a check *believing they have money in the bank account to cover it*, they may not have committed a crime. In fact, poor record keeping accounts for the majority of the bad checks written.

"Floating a check" is another example of a bad check where there may be no violation of criminal codes. "Floating" usually occurs just before payday. In this instance, the customer will write a check hoping to have money in their account before the check clears the bank. Many people practice this game of chance and never bounce a check. However, if you and the customer have the *same* bank, the check may clear much faster than the customer expected. Again, here the customer may have simply made an honest mistake; however, there are steps you can take to protect yourself from bad check writers.

Bad Checks

What To Do With Bad Checks

If you get a bounced check, the first thing to do is to call the bank upon which the check was drawn. The personnel at the bank can tell you if there are sufficient funds in the account to cover the check. If this is the case, re-deposit it. The chances are good that it will clear this time. If there are not sufficient funds in the account, call the customer. This is one of the best reasons to be certain that the customer's telephone number is written on the check. If you reach the customer, tell her the problem and ask for a specific date that funds will be available to cover the check. *Never* agree to hold the check until a certain date. If you do, the chances are that you have entered into a *verbal contract* and the police will not be able to help you. This is just like accepting a *post-dated check*. A post-dated check is a promise to pay *sometime in the future*. This is most definitely a contract which, if broken, is a civil matter which must be settled between you, the writer of the check, and the civil courts. Police only have authority in criminal matters and rarely, if ever, in civil cases. Remember, in most jurisdictions, for something to be a criminal act there must be *an intent to violate the law*. If you call the customer and discover that they have given you a bad, or non-existent phone number, call the police. By giving you a bad phone number, they have laid the groundwork for proving criminal intent.

If a check is returned stamped "account closed," the problem may be more acute. Call the bank to determine the date on which the account was closed and who took the action to close it. If you have failed to deposit the check for a week or more, the customer may assume that all his checks have cleared. Very often in divorce cases, one spouse will clean out the account without the other knowing it. If the other spouse closed the account, the one that wrote the check may be innocent of any wrongdoing. In such cases, call the check writer. This may take care of the matter. Be careful here, because both spouses may be involved and the one you talk to may play on your sympathies. You may be asked to hold the check for a couple of days so he or she can "get back on their feet." Do not agree to hold the check to any specific date, as this is a verbal contract thus requiring you to take action in small claims/civil court. If you are not satisfied that the check will be made good right away, turn it over to the police. If you are unable to contact the check writer or learn that you have been given false information (false address or phone number), take no further action other than to notify the police.

Bad Checks

Working With Police

Keep in mind that police departments are not collection agencies and when they become involved in a bad-check investigation, they do so for one reason: prosecution. They will expect that you are *willing to prosecute* if you wish them to take action. It is not unusual for the check writer to make the check good once they know the police are involved. Some police departments (actually, very few) will help businesses collect on bad checks. Most do not have the staff to do this and have learned that this type of service encourages businesses not to be as careful in the way in which they accept checks.

As a business owner, it is to your advantage to work with the police. You will find that most investigators (check forgery, burglary, robbery, etc.) will go out of their way to help you as long as you work with them in a cooperative and professional manner. The laws vary so much from state to state that it is most advantageous to have the bad check investigator on your side.

Laws involving bad checks can be very tricky; some of the reasons have already been covered. If you call a check investigator and ask him what can be done about a specific check problem, the most common response is "that depends." Then, you will be asked a series of questions to evaluate whether or not the case is criminal or civil and whether or not there are additional steps you need to take. Ensure that any follow-up requests for information are handled and reported back in a timely manner.

Many police agencies will have a "bad check" form that they will have you complete when submitting the check. Provide as much information as you can and follow through with any additional evidence that they may request. Such forms generally want to know who accepted the check (the employee's name); what identification was provided at the time the check was accepted; and whether or not the employee could identify the check writer if seen again. Also, there will be lines to identify the amount, check number, bank, account number, and name(s) on the check, as well as other pertinent information.

If the police, after review, tell you that your case is a civil case, you can file a small claims action. Check with your local courts to determine what dollar amount constitutes a small claims action (there is generally a limitation on the top end). The rules will vary from state to state. If you do go to small claims court, you will need to take the evidence with you—that being the original check (a photocopy is generally not acceptable due to the "best evidence" rule), a sales receipt from the register and, if you were fortunate enough to have had a video tape recorder (from a surveillance camera) running at the time, the video cassette. There's far more to small claims actions than just showing up and telling the judge what happened. The judge will want to hear from the witness(es) and view the evidence. In some cases, the only witness may be you. In the event the check was accepted by another employee, his or her testimony would be required, along with your testimony as to how you attempted to reconcile the matter out of court.

Protecting Yourself From The Bad Check Writer

One popular scheme that bad check writers use is where they actually have a legitimate-appearing checking account. This "checking account" was set up, however, utilizing stolen identification. The suspect will deposit a small amount of money into the account. For sake of discussion, let's say that the suspect deposited $200 into this account. The suspect will then go all over town writing checks, but always writing the check for just under the $200 balance. Any merchant verifying the check will be told by the bank that there is sufficient money in the account to cover the individual check in question. Later, the suspect will go to the bank and close out or drain the account, retrieving all or most of the original $200 deposit.

One way to beat this crook at his or her own game is to inflate the amount of the check, say by $50 or so, when seeking clearance from the bank. If the bank refuses to verify it at the slightly inflated amount, then move with caution. Unless the check writer is a good customer, you shouldn't accept the check for payment.

Of course, it is next to impossible to call the bank to verify every check received. Many customers would become angry at receiving this "third degree" treatment and might just say "never mind." Obviously, the business owner and the employees have to exercise good judgment. Somehow, you have to balance the good will that you wish to project with good business sense—nobody can afford to be stuck continually with bad checks.

Ask For Identification

So when do you call the bank to verify a check? If you do not know the person writing the check, *always* ask for identification. Accept only

Bad Checks

identification which bears the writer's picture and that has been issued by the state or federal government. It is quite easy to buy false identification, but not easy to duplicate the quality the government puts into a driver's license or a military identification card. So-called city or county identification cards can be purchased at any swap meet. As for Social Security cards—they're worthless. (One investigative reporter managed to obtain a social security card for his dog, just to show how easy it was. The scary thing was that he obtained it from the Social Security Administration, not from some table at a swap meet someplace.)

When you obtain a picture identification card, take time to look at it! Does the picture look like the person standing before you? Most bad check investigators have horror stories about clerks who accepted checks from persons who presented a photo identification bearing the photo of a different person, in some cases of a different race, and in others, of a different sex. Take time to look at the picture!

Of course, people do change. (Thankfully, we don't all look like our driver's license photos.) Their hair may not be the same color, style, or texture. Some grow beards or shave them off or gain or lose weight. But, if you're still in doubt, ask for a second identification and call the bank for verification. Remember, the bad check writer is counting on your hesitancy to call the bank and risk upsetting them, a "good" customer. They will use your hesitancy to their advantage and will manipulate the situation by being overly friendly and very talkative. Their goal is to keep you too busy to be suspicious and, even if so, too busy to take time to call the bank. Very often, they will write checks at the busiest time of the day, when they know you don't have the time to call the bank and wait on hold while it is verified.

Women outnumber the men two to one when it comes to writing bad checks. So, if you have a female customer that is very talkative at the busiest part of the day and she just doesn't look like the picture on her identification, you had better take the time to verify the check.

Checks with numbers from 1 to 200 should always get a second look. Checks with low numbers may indicate that the account has just been opened. This is not always the case, however, as many persons will ask that their first book of checks in their newly opened account, start with a high number, such as 1000 or so. This is done specifically for the purpose of not having to undergo the "third degree" when paying by check. Many banks will honor this request, because most of the time it is a legitimate request with no intent to use it as a tool to defraud.

Bad Checks

Make certain that you always write the check number on the receipt and never accept returned merchandise without it. If the check has not cleared the bank, no way are you going to give a cash refund.

Out-Of-State Checks

Only accept out-of-state checks if you are also willing to accept a great deal of risk. The police department, in most cases, will not spend several thousand dollars to bring a suspect back from another state on a minor case. It is also unlikely that a state would even extradite unless the losses are high enough to allow for a felony filing. How would you serve a person living in a neighboring state with court papers for a small claims action? These are the risks you must take if you accept out-of-state checks.

Third Party Checks

Third party checks are also highly dangerous. A third party check is when someone writes a check to another person and then that person brings you the check to pay for goods or simply to convert it into cash. What if the check is stolen? What if it were forged? You just might recover your money (if you're very lucky) through the courts. However, here you have to depend on other people to assist you at trial. If the person who received the check originally believed it was good, then it will be next to impossible to get that person to pay. Then, your only alternative is to go after the person who wrote the check. Now, you're depending upon the person who received the check to have exercised good judgment when accepting it.

If the check writer and the check passer are related, forget it. It is not unusual for drug addicts to forge their parents' checks, make the checks payable to themselves, and then attempt to cash them. Generally, when the check owner is contacted, they try to make it good—if they can. However, if they don't care anymore, you're faced with trying to get a parent to testify against a son or daughter. Third party checks are trouble—let your competitor cash them.

Stopping The "Pros"

Here is how the "real pros" do it. They first go to the county courthouse and tell the county clerk that they have lost their birth certificate. They use a name they copied off of a headstone out at the cemetery (the birth date is on the stone) and obtain a duplicate copy of their birth certificate. Next, they hit the local department of motor vehicles registry where they obtain a nice new driver's license bearing their picture and the name and date of birth of a deceased person and a fictitious address. To ensure that the picture is no more valid than the name, they wear a disguise of some sort—perhaps a wig, or maybe just some non-prescription, clear-lens glasses and a beard that they will shave off later. Using this shiny new driver's license as identification, they open up a checking account, requesting that the check numbers start with a number like 800 or something, so as not to confuse these checks with those of their prior banking institution and to avoid the nasty treatment one always get from merchants who recognize a new account because of the low numbering. Surely, it can't be this easy, right? Well, teenagers have discovered how easy it is: a large population of 17-year-old kids have driver's licenses showing them to be 25 or older—at least over 21 (the magic number for weekend fun).

How do you protect yourself? First, if the check is for a large amount of money, always insist on two forms of identification. One of the best things to ask for is a car registration in their name (but only ask for this after you have requested their license plate number). You can always tell them that your insurance company requires you to include this data on all large checks that you cash. After they tell you the license plate number (they will make it up, by the way), ask them for the registration. Don't be surprised if they say they rented the car they're driving (even that can be verified, to a certain extent). It would be rare that a professional will go to the trouble of having their car registered under a fictitious name. Even if they did, a car can be easy for police to trace. Sometimes, however, the car can be rented utilizing a fraudulent driver's license and paid for with a forged check. How do you protect yourself? Get to them before they get to you.

Other Preventive Measures

One of the best ways not to offend good customers and yet still scare off the pros is to have a check policy posted for all to read. On that policy statement, clearly state that cashing checks may require a vehicle registration. Another way is to indicate that checks over a certain amount require a thumbprint be placed upon the back of the check. Never state just what that amount is, otherwise the crooks will simply write a check for just under that amount—thus, no identifying thumbprint required. In order to carry out the policy, businesses may obtain Porelon® ink pads from police supply companies (and in some cases, police uniform shops). Another product on the market is an "inkless fingerprint" system, which uses a white tab (it comes in a

Bad Checks

handy dispenser) that has adhesive on the reverse side. This is affixed to the rear of the check. The check writer is then asked to touch their right thumb onto a chemical pad and then press their thumb onto this tab. The chemical is transferred via the friction ridges of the skin and reacts with the paper, developing a permanent print. Since this system does not use ink, there is no mess. The value of the fingerprint has increased dramatically now that many cities, counties and states have computerized fingerprint systems that can identify a suspect in a matter of minutes. The bad guys know this and also know that unless they can say the print was somehow forged, they are what cops call "dead players."

Check-alert services are available in most areas. Some are informal and run by the local chamber of commerce, while others are businesses that provide a service for a fee. The user can access the computerized files (or look through a print-out) to determine if the individual or the checking account number is "bad paper." You need to determine if the price for these services will be cost-effective for your needs. Quite frankly, there would be some value in simply acquiring a sign that says, "We verify all checks through Check-Alert" (or some other fictitious company). Valuable as such computerized files are, they are seldom updated quickly enough to catch the professional. They do work fairly well in tracking bad checks and checks stolen in your area.

Even if you don't use a check-alert service, don't fingerprint check writers and don't require vehicle registrations, a posted policy saying that you do may cause a bad paper hanger to make a U-turn and head out the door.

Always remember to write down the customer's phone number and driver's license number on each check. With the driver's license number, the police can obtain a photograph of the individual from the motor vehicles registry should it be necessary for a photographic line-up.

Chapter Seven

Counterfeit Cash

Counterfeit money is so abundant that the Bureau of Engraving and Printing has had to make its first major change in the look of our money since 1929 (and that wasn't such a good year). In 1990 alone, *over sixty-six million dollars in bogus bills were confiscated*. A lot of this "funny money" is coming in from other countries. Some of the third-world countries do little, if anything, to prevent the counterfeiting of U.S. dollars. Additionally, high-tech, color copiers can produce some really fine-looking currency (this is not a suggestion!). This chapter will provide you, the small business owner, with techniques to help you recognize counterfeit currency and steps to take in the event that you are delivered a counterfeit bill.

If you deposit a counterfeit bill that became part of your daily receipts, the bank will deduct the face amount of that bill from your total deposit. If you call the police, they will seize the bill and you're still out the face amount that the bill "represented." The best thing you can do, then, is to catch the bill at the register *before* it is accepted as payment for goods or services. However, because of the reasons mentioned, it is getting increasingly more difficult to spot counterfeit currency.

Since you handle currency every day, start paying attention to the feel of the paper. Counterfeit money does not have the same feel as real money does. The counterfeit money will feel slick, like regular writing paper, whereas real money has a feeling of texture. Some of the really poor counterfeit bills will be easy to spot because they look as though they passed through the washer. Sometimes the bills are so poor that the green will come off on your hands. These bills are usually passed to some young kid working behind a counter at a fast-food restaurant.

The next grade up in counterfeit bills is also easy to spot; however, it does take a little closer examination. Again, these bills will not feel right to you. When you take a closer look, you can see that the fine lines around the edges are broken or fuzzy looking. The lines in the President's portrait will look the same way. Sometimes one side of the bill will look better than the other. Another test involves holding the bill up to direct light. With a little luck and a keen eye, you should be able to see fine, red and green fibers in the paper of real U.S. currency. However, a lot of people have trouble seeing these small fibers. So, be sure to practice with some real money before you start trying to spot the counterfeit variety.

One mistake counterfeiters used to make is with the serial number of the bill. The serial number begins and ends with a letter. Real currency has the same letter that the serial number begins with in a seal which is located just above the serial number and to the left of the portrait (as you view the bill). Also, just above and below the letter (left side) is a number. This number should be equal to the placement in the alphabet that the first letter of the serial number represents. For example, if the first letter of the serial number is "A" then the number above and below it (to the left) should be a "1." If the letter is a "B" then the number will be a "2" and so on.

The Counterfeiting Process

One process of counterfeiting involves making photographic plates of the bill. First, they shoot a photograph of the dark green characteristics of the bill. A second photograph is taken aimed at recording only the light green characteristics. Since the serial number is in light green, this method makes it easier to change the serial numbers of bills. If you receive two bills (and either one or both do not feel right), check and see if the serial numbers are the same. Of course, if you have two serial numbers that are the same, you have two counterfeit bills.

The really good counterfeit money will still feel slick to the touch and will not have the red and green fibers embedded in the paper. Also, it will not have the mistakes in the serial numbers just mentioned. It looks *very* real. The people who make the stuff even wash it, so that it will not look new. The small lines around the edges and in the portrait will look as clear and distinct as they do in real money. These bills are the reason that the Treasury Department is making changes in the currency of the United States.

Counterfeit Bill Passers

The people who pass counterfeit bills are generally not the ones making the money. People who pass counterfeit money fall into two categories. The first consists of people who have had the money passed to them. They have no idea the money is counterfeit and are not committing a crime since they lack knowledge, and therefore, the intent necessary to commit a crime. There is one exception here, however. Occasionally, a drug dealer will unknowingly be given counterfeit money for drugs. These people will almost always have more than one or two bills on them—all counterfeit. The second type of person has purchased the counterfeit money. Counterfeit money, depending upon the quality, can be purchased for several cents to twenty-five cents on the dollar. This person will then go from business to business, making small purchases with large bills. Very often, they will try to hit a business when it is busy and will usually try to single out a young cashier. The purpose of course, is to trade their counterfeit money for the real stuff.

If you think someone has or is trying to pass a counterfeit bill—*call the police.* Even if the person who is passing the bill is innocent of any criminal intent, the police will still want to know from where the bill was obtained. If the person has more than one or two counterfeit bills, the police (and eventually the Secret Service) will be very interested in talking with this individual. The law can be very tricky here, so unless you know what you are doing, a citizen's arrest is not a good idea. Just try to stall the person until the police arrive. But don't hand over any of the money or any of the merchandise. If necessary, tell the person that you suspect the bill is counterfeit and that you have called the police to check it out. You may even want to suggest, "Oh, we get them all the time." Whether such is true or not, this may help to reduce the person's fear and hopefully keep them there. It may be valuable, both to keep a good customer and to entrap a crook, to say something along the lines of, "We don't think you've done anything wrong—you're a victim in this, too."

The Latest Detection Techniques

A new item on the market is a chemical pen that detects counterfeit money by testing the the paper itself using a chemical reaction method. Using the pen, a stripe is placed onto the suspect bill. If it is in fact U.S.

Counterfeit Cash

currency paper, the stripe will turn a pale yellow. If it is a counterfeit bill, the stripe will turn a chocolate brown. This is probably the simplest method of checking the authenticity of a suspect bill. The pens come in a set of six for about $25.00. Again, check with a police supply company (criminalistics supply catalogue) if you think you need one. Additionally, one can purchase a *linen tester* (essentially, a small magnifying glass that renders fibers quite visible) at the local camera store. It is used by photographers to closely examine film negatives for flaws. This will allow the fibers within the bill to be more easily seen (if they are there). There are also black light devices on the market that can help detect counterfeit money. However, if counterfeit money is not a problem for you, save your money.

When the Treasury Department modifies our currency, the modified money should make things a lot easier. These bills will contain a polyester filament imprint with extremely small lettering. This will be located on the left side of the bill and will run from the top to bottom through the area of the first letter of the serial number. Visible only if held up to direct light, the thread will read "USA 100" for a $100 bill. The new currency will also contain micro-engraving around the portrait. In extremely fine print is "The United States of America." This is repeated around the rim of the portrait. When you view the bill, the micro printing will look like a solid line surrounding the portrait. With magnification, the words will be visible. The Treasury will start with $100 bills and work down to $50 bills. This micro-engraving, as well as the security thread, will eventually be placed on all denominations of paper currency, with the possible exception of the $1 bill. The major benefit is that the two new security measures described cannot be reproduced—even by the best copy machine.

The following material was reproduced from a brochure provided by the U.S. Secret Service and is available to the public through the U.S. Secret Service or through the U.S. Government Printing Office:

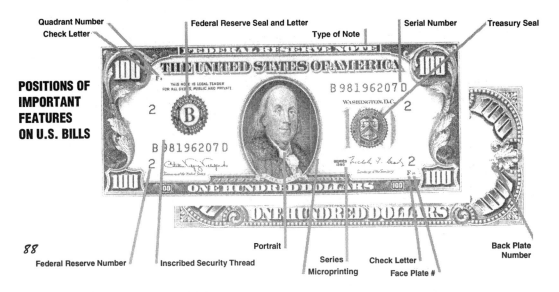

POSITIONS OF IMPORTANT FEATURES ON U.S. BILLS

Counterfeit Cash

DESIGN FEATURES FOR NEWLY ISSUED CURRENCY

Due to increases in color copier technology, two new security features are being added to U.S. currency. These new features will appear first in Series 1990 $50 and $100 Federal Reserve Notes. Additional denominations will be gradually phased in. Existing currency and the new series will co-circulate until existing currency is withdrawn at the Federal Reserve banks and branches. Withdrawal will be based on normal wear.

Inscribed Security Thread

A clear, inscribed polyester thread has been incorporated into the paper of genuine currency. The thread is embedded in the paper and runs vertically through the clear field to the left of the Federal Reserve seal on all notes except the $1 denomination. If it is decided to use the thread in the $1 denomination, it will be located between the Federal Reserve seal and the portrait.

Printed on the thread is a denomination identifier. On $20 denominations and lower, the security thread has "USA" followed by the written denomination. For example, "USA TWENTY USA TWENTY" is repeated along the entire length of the thread. Higher denominations have "USA" plus the numerical value, such as "USA 50 USA 50" repeated along the entire length of the thread. The inscriptions are printed so that they can be read from either the face or the back of the note. The thread and the printing can only be seen by holding the note up to a light source.

Microprinting

Concurrent with the addition of the security thread, a line of microprinting appears on the rim of the portrait on $50 and $100 denominations, beginning with Series 1990. The words "THE UNITED STATES OF AMERICA" are repeated along the sides of the portrait. As with the new security thread, the microprinting will also be gradually phased in on all denominations, with the possible exception of the $1 denomination.

To the naked eye, the microprinting appears as little more than a solid line and can only be read by using magnification. Neither of the new security features can be accurately reproduced by an office machine copier.

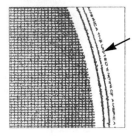

HOW TO DETECT COUNTERFEIT MONEY

Genuine money is made by the Government's master craftsmen who use engraved plates and printing equipment designed for that purpose. Most counterfeiters use a photo-mechanical or "offset" method to make a printing plate from a photograph of a genuine note.

You can help guard against the threat from counterfeiters by becoming more familiar with United States money.

Look at the money you receive. Compare a suspect note with a genuine note of the same denomination and series, paying attention to the quality of printing and paper characteristics. **Look for differences, not similarities.**

Portrait

The genuine portrait appears lifelike and stands out distinctly from the fine screen-like background. The counterfeit portrait is usually lifeless and flat. Details merge into the background which is often too dark or mottled.

Federal Reserve and Treasury Seals

On a genuine bill, the sawtooth points of the Federal Reserve and Treasury seals are clear, distinct, and sharp. The counterfeit seals may have uneven, blunt, or broken sawtooth points.

Counterfeit Cash

COUNTERFEIT COINS

Genuine coins are struck (stamped out) by special machinery. Most counterfeit coins are made by pouring liquid metal into molds or dies. This procedure often leaves die marks, such as cracks or pimples of metal on the counterfeit coin.

Today counterfeit coins are made primarily to simulate rare coins which are of value to collectors. Sometimes this is done by altering genuine coins to increase their numismatic value.

The most common changes are the removal, addition, or alteration of the coin's date or mint marks.

If you suspect you are in possession of a counterfeit or altered coin, compare it with a genuine one of the same value.

If it is above 5 cents in value, it should have corrugated outer edges, referred to as "reeding." Reeding on genuine coins is even and distinct. The counterfeit coin's reeding may be uneven, crooked, or missing altogether.

IF YOU RECEIVE A COUNTERFEIT:

1. Do not return it to the passer.
2. Delay the passer if possible.
3. Observe the passer's description, as well as that of any companions, and the license numbers of any vehicles used.
4. Telephone your local police department or the United States Secret Service. These numbers can be found on the inside front page of your local telephone directory.
5. Write your initials and the date on a blank portion of the suspect note.
6. Do not handle the note. Carefully place it in a protective covering, such as an envelope.
7. Surrender the note or coin only to a properly identified police officer or U.S. Secret Service agent.

WHEN MONEY IS DAMAGED OR WEARS OUT

Even though United States currency is strong and durable, it does wear out with constant handling.

All currency in circulation is routinely deposited in Federal Reserve Banks by commercial banks. Worn bills are destroyed by Federal Reserve banks during ordinary currency processing. The destroyed bills are replaced by new currency provided by the Bureau of Engraving and Printing. The bill most commonly replaced is the $1 denomination. There are over 4 billion $1 bills in circulation, and the life expectancy of each is approximately 18 months. Since larger denominations are handled less, they last longer.

When a bill is partially destroyed, the Treasury Department will replace it if clearly more than half of the original note remains. Fragments of mutilated currency which are not clearly more than one half of the original whole note may be exchanged only if the Director of the Bureau of Engraving and Printing is satisfied by the evidence presented that the missing portions have been totally destroyed.

Damaged or mutilated bills should be taken to a bank for redemption. When partially destroyed currency is of questionable value, the fragments should be sent by registered mail to the Department of the Treasury, Bureau of Engraving and Printing, OCS/BEPA, Room 344, P.O. Box 37048, Washington, D.C. 20013.

Chapter Eight

Counterfeit Products

The counterfeiting of products has reached epidemic proportions. In recent years, everything from baseball cards to expensive perfumes has been counterfeited. Products with designer names have been and *are* the favorite targets of counterfeiters. Most of these counterfeit products are manufactured in foreign countries, where they are sold at a much lower cost than the brand-name merchandise. This not only hurts the companies with the copyrights or trademarks, but ultimately victimizes the consumers. And of course, if your competitor is selling these items at a lower cost, it is going to hurt your business.

Counterfeit products are infringements of trademarks and copyrights. Generally, this means that if you're knowingly selling counterfeit goods, you are in violation of federal laws. Additionally, you're committing fraud, which is a violation of state law. Even if you don't know you're selling counterfeit goods, the police or federal marshals can seize these products right off the shelf and right out of the storeroom. It shouldn't come as a surprise that you're not going to be reimbursed for the items seized and you are correct in assuming that you won't get the items back.

Protecting Yourself Against The Counterfeit Product

How can you avoid becoming a victim? First of all, know the products you sell. Know what the products look like and what markings are pre-

Counterfeit Products

sent. If you receive a shipment from your supplier of a particular product and it doesn't look right, question it.

Know how the products are packaged. Counterfeit products are generally packaged with less care than the real thing. The packaging may be a better tip-off than the product itself as to whether or not you're looking at a counterfeit product.

Your best bet is to deal with suppliers that you know and trust. If someone approaches you with a deal that sounds too good to be true—it probably is. The usual line is, "We buy in such large quantities that we can afford to sell them to you for much less." (This certainly would not explain why the showroom is the individual's car trunk—that's only one step from the inside of a trench coat!) Before you buy from any supplier offering a "too good to be true" price, call the manufacturer or owner of the trademark and ask them if the person with whom you are dealing is an authorized supplier.

Most companies can supply you with pamphlets that will show you how to identify their products. These pamphlets will contain information on who to contact for more information and what to do if you discover a counterfeit. The larger companies have a full-time security staff that will investigate any infringement of their copyrights or trademarks. If in doubt of whom to call, contact your local police or the nearest office of the Federal Bureau of Investigation.

Specific items to watch out for include high-cost jewelry items with easily recognizable name brands, such as Rolex. The product-type may also include clothing items that may not be expensive individually, but collectively, would result in a major loss to the defrauded manufacturer and a boon to the defrauding crook. In today's society, especially among the young, having "name brand" clothing items is of utmost importance. The brand-name placed on the clothing represents its value. Some kids have gone as far as sewing the labels from more expensive clothing items on their less expensive clothing. (Some adults have done this, as well.)

The Levi-Strauss Company is presently battling a major importation of its highly famous and bestselling 501-style jeans. The jeans are being manufactured and imported from China. Levi-Strauss Company officials say that they have never seen anything occurring on this scale, this scope, this price point, and of this quality. They describe it as the worst counterfeiting the company has seen in its 140-year history.

Chapter Nine

Charge Accounts

If you allow people to charge merchandise at your business, several problems can develop. All charge accounts are contracts and any problems with payment, other than a scenario involving out-and-out forgery, are civil in nature and not a police matter.

One major problem that can occur is when a customer's employee is allowed to charge. It is not unusual for this employee to take advantage of the arrangement and charge a little extra for themselves. Certainly, the customer owning the account can fire the employee or even try to charge the employee with theft. Where the real problem occurs, however, is when the dollar amount really gets out of line and the account owner refuses to pay it. They will argue in court that you and your employees did not act reasonably and allowed the unauthorized charges to occur over a long period of time without making any attempt to notify them. In other words, you should have known better. The sad thing is that they just might pull it off. (The same thing can occur when parents allow their children to charge on their accounts.)

One excellent way to protect yourself from this possibility is to utilize a written agreement for all charge accounts. The agreement should include *who can sign on the account* and *the dollar amount that can be*

Charge Accounts

charged by each person. The agreement should also include *the signature of each person that can utilize the account.* You should check this signature card every time you are not familiar with a person who represents himself as an employee wishing to charge items to the "boss's account." This will prevent a forgery. When in doubt, call the account owner and verify the purchase.

Opening Charge Accounts

When a new customer walks through the door (or for that matter, even a customer who has done business with you for some time) and asks to open up a charge account, be aware that you are entering into an agreement that, if handled incorrectly, could cost you a great deal of money.

A business cannot make money unless it can sell a product or a service. Therefore, any new business is always welcome. Since new business does not always turn out to be good business, you should take steps to protect yourself. How careful you wish to be in the extension of credit is directly related to the amount of money involved and how much you are willing to risk losing.

The foundation for any monthly billing program should be a written and signed agreement. The contract should spell out exactly who is allowed to charge on the account, the limit of any single charge (without a signed purchase order or other authoritzation), and the total limit of charged goods and services for any calendar month (unless other arrangements are made in advance). The contract should identify the date by which payment is due and what steps will be taken if payment is not received. This would include, among other things, a percentage penalty on the unpaid balance and the loss of charging privileges until the outstanding balance is paid.

Even though your store may be small, you still need to be using a contract. Being a smaller operation, you can less afford to have someone default on a major bill. You can sell the formal contract to the customer as protection, too (which it surely is). It is suggested that you consult an attorney to assist you in drawing up a contract which provides maximum protection to you in the event of a default in payment. Another option is to utilize standardized forms (often available through your business associations.) However, be sure that the wording of the contract protects you sufficiently.

Dealing Effectively With Law Enforcement

If a new customer, be it another business or an individual, wishes to initiate a charge account, have them first complete a credit application. This should always include a waiver allowing you to run a credit check. Look at the credit history and check the references given. Use your network of fellow business owners to find out if the new customer is a good credit risk. It is possible that the only reason you are being asked to extend credit is because the last place said "no more!" If you own a paint store, check with a few of the other paint stores in town to see if they have extended credit to the individual (or business) in the past. The bottom line is: Were the bills paid on time? If the individual or business representative claims to have been in business for several years, dig out a few old telephone books and see if they were listed. Check with your local Chamber of Commerce to see if the business is a member and when they joined. What is their status with the Chamber? Drive by the "business" to make sure that it exists—some business addresses are little more than a post office box service!

The key here is to not handle something as important as the extension of credit with little concern for failure. It may seem somewhat "bureaucratic" to have someone you know or a company with whom you have done business complete thorough credit applications. For the same reason, it is easy to fall into the trap of failing to do background checks prior to authorizing credit. You simply cannot afford to do business in this way any longer. Some business owners are quick to file bankruptcy rather than downsize and stick it out. You could easily be stuck with a major loss.

Even if the company is unquestionably solvent or the individual a pillar of the community, use a contract and treat the extension of credit as a business transaction requiring documentation and review prior to approval. Major companies have been knocked to their knees by a single individual's seemingly sudden addiction to cocaine (which, in turn, explains the massive embezzlements and loss of major contracts). Where addiction to any drug is concerned, there is a paradigm shift—the priorities all change and what was agreed upon prior is no longer recognized.

There is also the reality that some businesses are created simply for the purpose of establishing credit, maximizing-out the credit limits, and then filing bankruptcy. And of course, it's all done through incorporation, where there is nobody to sue. They just move on to do it again, elsewhere. This is one of the reasons you should always drive by the given address of the new business—it may be a mailing service or a barber shop where your customer "picks up his mail."

Charge Accounts

Handling Late Payments

When a local individual or business is late in delivering a payment, most creditors find themselves in somewhat of a quandary. They desire payment and don't want to risk losing a great deal of money, but they also do not want to risk damaging what has been a good business relationship. Most businesses start soft and easy and then escalate if necessary. For example, when failing to receive a payment which is due on the first, you may consider waiting until the tenth day of the month to give the business a little working room. Some businesses hold onto their funds until the very last day possible and then disburse them. By doing so, they earn the maximum interest on money in their accounts. Sometimes a payment may arrive a day or two late. Other explanations can be found in unforeseen events, clerical errors, or the delivery speed of the mail.

If no payment is received by the tenth of the month, a letter is sent to the individual whose name appears on the account informing them that (1) their payment was due on the first of the month, as per the signed agreement; (2) that it has not been received; and (3) might the payment be made by _____? Generally, the request is closed with "If this bill has already been paid, please disregard this notice. Thank you for your business."

If no payment is received by the date given, a phone call to the business or individual is made. The attitude is one of "how can we work this out?" Businesses often find themselves with a lot of cash on the books, but little in their accounts until THEY get paid. There may be an explanation that is understandable. Unless a less-than-accommodating attitude is encountered, there is little to be gained by demanding immediate payment "or else!" You're still hoping that they'll come through. (Keep in mind that it is still better to get part of your money back than none at all.) If you're getting the runaround, you can diplomatically state that you really don't want to see this outstanding debt go to collections or into civil court, "where nobody really wins."

If the phone call fails to generate the payment, the next step is to send a registered letter to the individual or head of the business (as identified in the credit application). This letter will detail the payment requirements (as set forth in the contract that was signed by them), the amount of the outstanding debt, and the collection attempts that have been made to date (letters, phone calls, etc.). The letter advises that a payment must be received by a specified date (or other arrangements, suitable and agreed upon by both parties, must be reached). Most people know that receiving a registered letter regarding funds due is generally a precursor to the account being turned over to a collection agency or the initiation of a civil suit. A registered letter is used because it must be signed for by the person to whom it is directed, and it serves as evidence of a demand for payment.

Seeking Justice

It is very difficult (if not impossible) to be able to prosecute credit or charge account violations as criminal matters. The extension of credit is a civil agreement between the business and the customer and, with few exceptions, cannot involve the police. Since you will generally have to handle violations of these agreements through civil court actions, your ability to identify the individual involved and produce the signed documents is of critical importance.

If it can be shown that there was fraudulent intent, the prosecutorial agency in your jurisdiction may be willing to take a look at the case. For example, one business was bilked out of $500 by a slick salesman who promised $1,500 worth of advertising on radio and in the print media. It never came forth. The check the business owner had written was cashed by the salesman. He simply added his name to the check—after the name of the company—and the bank cashed it. (This is a good reason why you should draw a line to the end of the "pay to the order of" block just as you habitually do on the line where the dollar amount is indicated—so nobody can add their name on.) Though this was certainly not a charge account, it was a civil agreement between the business and the salesman who represented another firm—much like the agreements you make when you extend credit. The forgery of the check and the salesman's total inability to provide the services offered for the amount charged were sufficient to allow entry of law enforcement and the prosecutor's office. Still, the victim business never saw their $500 and they received no advertisements, just headaches.

Trial Use Requests and Rental Equipment

Be wary of individuals who ask to take items from your business for "trial use periods." This is not much different than allowing someone to take an item or use a service for which they agree to pay for later. This is often the beginning of a major fraud. In one case, a nicely dressed gentleman from "out of town" entered a small office equipment business. He identified himself as a representative of an account-

Charge Accounts

ing firm in another city. He found a specific ten-key calculator that he liked very much. However, since he would not be the one using the product, he asked if it would be possible to take one for a trial use "by one of the girls." If the office staff liked the unit as well as he, the unit would be purchased. He informed the owner that he might be buying as many as twenty of the units. To the small town dealer, selling twenty units in one fell swoop was a rare opportunity. The three piece suit (and the luxury automobile that he drove) evidently convinced this business owner that the customer was legitimate, because he ultimately agreed to let the gentleman take the total number of units he needed (twenty) with an agreement that he try them for thirty days and then either purchase them or return them. The agreement, by the way, was nothing more than a sales slip bearing a generic description of the property and the signature of the customer. On the base of the sales slip, the owner wrote "thirty day trial."

The customer in this case was a con man. The only proof the victim had that this customer truly owned a business was that he had a business card! Truth was, he didn't even know if the man presenting the card was, in fact, the man whose name appeared on it. The con man never paid for the calculators nor did he return them. What appeared to be the "sale of the century" for this small town dealer nearly became his undoing. Had he called the number on the business card prior to handing over the units, he would have found that the number was no longer in service. Had the victim checked with the Chamber of Commerce in the city where the business was located, he would have found that the accounting firm he claimed to represent was now a vacant building scheduled for the wrecker's ball. Are you giving out credit with no more protection than this?

Rental businesses are constantly the victims of rental agreement violations. All too often they are willing to accept most anything as "identification" and are willing to send the customer out of the business yard with a $1,000 piece of machinery retaining only a $75.00 deposit and a signature as guarantee that the equipment will be returned. In many cases, the property is not returned and the business owners quickly find out they can't even prove to whom they gave this expensive piece of property and don't know how to get in touch with him.

If rental businesses would take just a moment to step into the back room and call the residence and/or business phone number given by the renter on the agreement, they could verify the address and name of the renter. At the renter on the agreement, they could verify the name and address of the renter. At least if the renter failed to return the property, the rental business would have something to go on. (Why not place the renter's thumbprint on the contract?)

In many states, once the agreement is signed and subsequently violated, the business need only to send the individual a registered letter demanding that the property be returned and all fees paid by a specified date. Once the date has passed, many prosecutorial agencies will view the failure to respond to the registered letter as evidence of intent to steal, allowing the case to be investigated by the police and prosecuted as a criminal, rather than civil, matter (provided the business obtained accurate identification of the individual who rented the property).

In Conclusion

Bottom line—do not treat any business transaction informally, unless you are willing to risk the loss of property and/or money. Law enforcement can do little to assist you and you can do little to help yourself if you fail to obtain valid, verified identification, addresses, etc., of those to whom you extend credit or property. Both civil and criminal courts demand that business dealings be documented, with each party's responsibilities clearly stated and the agreement signed by both parties.

Chapter Ten

Dealing Effectively With Law Enforcement

It might seem a bit on the foolish side to include within this guide a section on how to deal more effectively with law enforcement. Also, to suggest that they must be dealt with differently would also suggest that they "aren't like normal people." Perhaps to some degree this is true. What one must remember, however, is that law enforcement officers belong to law enforcement agencies, which are major bureaucracies (complete with all of the problems associated with bureaucracies). This bureaucratic nightmare is compounded by the rank-and-file structure of such organizations—where the parameters of responsibility and authority are well-defined and there is always someone higher up to which an appeal may be filed. Despite the bureaucratic nature of the law enforcement profession, somehow they manage to pull off some amazing things. Crooks do go to jail for their crimes as a direct result of an officer's quality police work and tenacity. The goal of this section is to familiarize you with a few realities of police work so that you might operate from a more informed position and, therefore, more effectively.

The Life Of The Police Officer

Law enforcement officers are not like normal people. This statement is not meant to attack the personalities of the men and women in blue.

Dealing Effectively With Law Enforcement

Rather, it is the best and most succinct way to get the point across that law enforcement officers are *different*—not abnormal. The question is, *why are they different?*

Well, there's shift work to start with. What constitutes a day shift, a swing shift (evenings), or a graveyard shift (overnight) may differ with the agency. The one commonalty is that nobody really likes shift work and there are probably as many officers who hate working day shifts as there are officers who hate working "graves." Day shifts are spent doing little more than working traffic collisions involving persons who left too late for work and cannot let rain or fog slow them down (you feel like you're an insurance company representative, not a cop). Graveyard shifts are one family disturbance after another and most everyone you contact is either drunk, or under the influence of drugs. You cannot understand why two adults, married to each other, have to have a cop in their living room to referee a fight over who flirted with who at the bowling alley. Evening or "swing" shift is generally so busy that the dinner break you were supposed to get at 7 P.M. doesn't come until 9 P.M.—if it comes at all. Regardless of which shift you work, a twenty-four hour business means that you're never closed. You work Christmas, New Year's, the Fourth of July, your birthday, your kids' birthdays, your anniversary and you don't get to watch your kid play baseball or act in the school play.

Then there's court. You get off of a graveyard shift at 7 A.M. ready for bed. Unfortunately, a subpoena for a year-old petty theft case will keep you from your slumber. You wait in court most of the day for your case to come up, only to be told that the suspect pleaded guilty or that the case has been continued until next week. You go home mad and manage to get four hours of sleep before it is time to go back to work. The only thing worse than court after a graveyard shift is being "on call" for court on your days off. You can't go anywhere. You can't even mow the lawn for fear of missing the phone call that you're needed in court.

Police officers must have absolute emotional flexibility. You will go to a call involving a psychiatric patient that went crazy at home and are only able to subdue and restrain him after a fight *and* after ripping your $130 pair of uniform pants. Fifteen minutes later, you're taking a report from a hysterical lady who's reporting that her fourteen-year-old daughter is an hour late getting home from the roller rink. She may later call and complain to the supervisor that you were "insensitive" to her trauma.

Often, you're the one to deliver bad news. You get to go and tell people that someone they care about has died. You get to tell them that their loved one has been involved in a critical-injury collision. You get to tell them that you picked up their thirteen-year-old daughter drunk, naked, and in a somewhat "compromising" position in the back seat of a car. You get to tell the victim of a crime that there is simply no way to prove their allegations and that, yes, the perpetrator did get away with it. You may even get to tell them that the terrible thing that was done to them isn't even against the law in this state.

Sometimes you take on the role of a verbal punching bag. A homeowner gets ripped off in a burglary (where entry was made by simply walking in their unlocked back door) and in a state of shock, disbelief and anger, you are treated as if you recruited the crook and told him to commit the theft. They challenge your reason for existence and suggest that, if you weren't "sitting in the coffee shop flirting with waitresses," the crime might not have happened (even though you can't remember the last time you had time to stop for a cup of coffee). And of course, they've NEVER seen a cop car drive down their street—ever. The professional police officer understands that this is a venting process and empathizes with the victim. Police officers would be angry, too, if they were burglarized.

And then some lady calls screaming for help in trying to control her 16-year-old, long-haired, dope-smoking, heavy-metal son who threw a fit when mom said he couldn't have sex with his girlfriend in HIS bedroom and then started destroying everything. And when at her request you take this young'un into custody and have to roll around on the floor with him, before you can get the cuffs on, she screams, "Don't hurt my baby!" Later, she will pay for his attorney to beat the charges filed by that "no-good cop."

The Men And Women In Blue

If things are this bad—why would anyone be a cop? The answer is: "because the good outweighs the bad." When you help extricate an injured child from a mangled crash scene and buy that kid a few toys out of your own pocket, you get a feeling that cannot be duplicated anywhere else. When you pull out the three-year-old girl who rode her tricycle too close to the edge of the decking and ended up in the deep end of the pool, resuscitate her and give her life back to her, you get a rush that cannot be duplicated anywhere else. And when you nail a low-life that has pulled a string of armed robberies in which he severely beat the store clerks, you get a feeling of satisfaction that you couldn't get being a car salesman or a banker or a mail carrier. Then, there's the rapists and child molesters, and those who steal from the very young, the very old, and the very ill—defenseless human beings.

Dealing Effectively With Law Enforcement

Most cops are in the profession because they still cling to the values that made America great. They believe that if you work hard, you will be rewarded. They believe that most people are moral, ethical, and responsible. They believe that a third-grader ought not to have his bike stolen while he's playing baseball with his buddies; that elderly ladies ought to be able to go shopping without having their purses snatched and their hips broken; and that one ought to be able to go to dinner and a movie without being robbed and shot or stabbed. They are reasonable enough to know that individually, they will not change the world, nor rid it of its ills and its perversions. They just want to be part of the movement that is trying. Many times, they get personally involved and share the frustration, the fear, the hurt, and the anger of the victims—victims who remind them of their mothers and fathers, sons and daughters, and friends.

So if you summon the police regarding a customer that shoplifted a five-dollar item and fled, and the officer seems somewhat unmoved by the event, remember where he's been and what he's had to do. And if he gets a little defensive when you vent your frustrations about how long it took for him to arrive, keep in mind that your call was only one of many he has worked and will work during that shift. Cops receive interpersonal skills training that is designed to increase their effectiveness in dealing with the public. They are taught to show empathy for victims of crime and to give your call the quality attention it deserves. For the most part, they do a pretty good job of it. They are, however, recruited from the same human race that you came from and they will sometimes make mistakes.

Things You Can Do

How does one work more efficiently with law enforcement? Perhaps it is easier to discuss some things NOT to do. For example, do not repeatedly set off your burglar alarm trying to open or close the store. In most cases, officers will have to respond to your store even if you call and say the alarm is false (because burglars have been known to call, as well). Don't use your hold-up alarm to report a traffic collision in front of your business—use the phone. Otherwise, the agency may have to send two officers to the collision and two officers to the robbery alarm. When you are the victim of a crime and summon police for a report, don't try to conduct business as usual, waiting on customer after customer and answering the investigating officer's questions in between. If it's important enough to call police about, it should be important enough to close for five minutes or seek relief to allow you to make the

Dealing Effectively With Law Enforcement

report. And don't call an officer to the scene, allow him to conduct a lengthy investigation, and then drop the bombshell, saying: "I don't want to prosecute anyone; I just want my stuff back." Cops prepare cases for prosecution; they are not collection agents.

Don't expect the unreasonable. The officers who arrive at your store to take your theft report will not spend the rest of their shift following up leads and interrogating possible suspects. They will write the best report that they can within the time allotted and forward it on to detectives. They will leave your establishment and head to the location of the next call, where there is another victim desirous of police intervention. If you have been ripped off and nobody can identify the suspect, if seen again, and nobody was able to see the license number of the car the suspect fled in and you couldn't positively identify your property if it were found—don't expect magic. The badge pinned on the officer's uniform will not make up for the obvious deficiencies in the case. And when your alarm is broken again and you have to leave it off, don't think that notifying the police department will put a police unit at your front door until the sun rises the next morning (although cops generally do make passing checks of businesses in such situations).

It is not intended to suggest that any member of the public, who is served by a law enforcement agency, must tolerate rude or unprofessional conduct by a police officer. Quite the contrary, for it is this type of officer that stands in the way of success. Nor is it meant to cause you to believe that law enforcement today is so ineffective, that utilizing them is wasted effort. What is offered here is *insight*—the opportunity for you to share some of the frustrations felt by today's law enforcement officers and the knowledge that no law enforcement agency anywhere is totally capable of preventing even *most* crime. (If law enforcement believed that they could, they wouldn't be spending millions of dollars annually on crime prevention and citizen education programs!)

So what does one do with this knowledge? With the understanding that you now possess, you are in a position to deal more effectively with law enforcement officers. Realizing that a petty theft (shoplift) of twenty dollars worth of merchandise just doesn't rate high on the criminality scale, you should downplay its importance to the officer. Offer your apologies for having to call them out on such a minor crime. If the officer is worth his salt, he'll probably respond by telling you that investigating such cases is his job and that shoplifting, cumulatively, is a major crime for which we all pay higher prices and that most of the time shoplifters have warrants for their arrest or are under the influence of drugs. Be adamant in your desire to prosecute the offender.

This tells the officer that you see his time and effort as deserving of some sort of equity. Certainly, if the "crook" is a young child who lifted his favorite candy bar or if there simply isn't sufficient evidence to prove the crime, you may want to ask for the officer's evaluation of what would be a reasonable response.

If you're fighting a vandalism problem (or any other problem) and really need to have a patrol unit swing by your business, occasionally try prefacing any request for a patrol car to pass by with: "I know you guys are really busy handling much more important things, but I would really appreciate it if a patrol car could drive by my business and...etc." Chances are, you'll have the cops sitting in front of your business writing their reports.

Cops generally feel like their efforts are unappreciated. Often times, they even question whether or not what they're doing means anything to anybody other than themselves. Like all people, cops like to be recognized when they do good work. Nothing does more for a cop's ego and self-esteem than walking into the station and finding a letter posted on the bulletin board, written by a business owner and addressed to the chief of police or sheriff, in which the writer praises the officers, by name, for their efforts. The fact that the citizen took the time to write tells these officers that what they are doing is worthwhile.

But what about the other side of the coin? What about the officer that fails to do his job—the officer that is lazy, or unconcerned, or just downright unprofessional? To allow him to continue on, unnoticed, is every bit as bad as allowing an officer who busts his behind to go unrecognized—perhaps, it is worse.

If discussion with the officer yields no results, it is reasonable to ask that a supervisor be called to the scene. In some cases, it may be better to discontinue the contact with the officer, allow him to leave and then recontact the agency to request that a supervisor call or come by your business. In your conversation with the supervisor, make it clear that your goal is to seek a remedy to the problem at hand, not to seek punishment for the officer. (The supervisor will generally see to it that the matter is dealt with.) In some cases, however, the officer's conduct may be so unprofessional, that you may desire to lodge a formal complaint. It is your right to do so and your voice will be heard. Closely evaluate your problem, however, before pushing the issue further. Is the agency representative being unprofessional or are your expectations unreasonable? As long as cops are recruited from the human race, there will be some that slip through the testing who should have been culled out.

Bad cops are bad for the citizens, bad for the law enforcement profession, and bad for the department.

If, after speaking with the supervisor (usually of sergeant's rank) and you still feel your voice hasn't been heard, you may elect to move another "rung up the ladder," and seek to speak with the sergeant's supervisor (generally a Lieutenant). Next would be a captain and then the chief of police or sheriff.

It's said that "the squeaky wheel gets greased first." While this is not the recommended method, one can threaten to contact the city council or county board of supervisors (or any other citizen's board with the power to hire and fire) or the press. Generally, you will be guaranteed to get the law enforcement agency's full attention, but they still may be unwilling or perhaps unable to provide the assistance you seek.

By resorting to these steps, you may alienate those whose assistance you are trying to acquire. If your complaint is justified, however, you will not alienate any professional officer—they don't like cops who don't pull their own weight or whose actions or inaction make the whole profession look bad.

Many agencies have "ombudsmen" or community relations officers who "know the ropes" and can cut through all of the red tape in order to get the action needed for your problem.

In Conclusion

Keep in mind that law enforcement agencies are "service" organizations who truly desire to keep the customer happy. In many cases, however, law enforcement organizations are restrained by policies, local regulations, state and federal laws, staffing, and calls for service—restraints that may prevent them from taking the action they would like to take. In short, quite often, it is the agency's inability to provide requested services which creates the problem, not their lack of desire to be of assistance.

Chapter Eleven

Firearms In The Business: Requirements And Limitations

Preparing Yourself

You may personally believe that all armed robbers should become organ donors. However, any plans you may have on shooting it out with the first armed robber that walks through the door of your business should be carefully considered.

Understand The Law

First of all, KNOW the laws pertaining to the use of deadly force in your state. If you've moved or been transferred, do not assume that they are the same as in the state from which you have come—*they can differ greatly*. Most, if not all, states allow you to defend yourself—it is the specific circumstances that vary from state to state. Can you shoot the suspect as he is leaving? In most states the answer is "no." What if you chase the suspect out of your business with your gun in hand just as the police arrive? You are now the target of the police. And don't think that your uniform smock or name tag represents some magical "King's X." Crooks have been known to exchange shirts with clerks to assist them in their departure. Don't put a gun in your store until you are fluent with the laws governing its use. And if you intend to leave it for other employees, they must also be well-versed in the requirements of the law.

Become Proficient With Your Weapon

If you have a gun, learn how to use it. Take a class at the local community college or something offered by the local gun club. It is surprising how many gun owners know absolutely nothing about how guns work. It might as well be a toy. Any training you obtain for yourself should be imparted to all employees, as well. You are responsible, to a great extent, for what they do. Failure to obtain training and failure to provide for training of your employees can be considered negligence, for which *you* can be liable.

If you are going to get a gun, get one with some "knock down" power. The object is not to kill the robber, but to stop him. Most people are not proficient enough to try to "wing" an armed suspect. The shooting of guns from suspects' hands and the disabling of crooks by shooting them in the arm occur mostly in the movies. When you're "pumped" (breathing hard and scared to death), you'll be lucky not to shoot your own customers. You shoot to stop, not to kill—but recognize that you may kill in the process. Do not shoot anyone that you do not have the legal and moral right to kill. Remember too, that you are responsible for where that slug goes. If you miss and the slug goes through a window and strikes a baby sleeping in a car seat outside the store, you alone are responsible for the proximate cause of that child's death. There's nothing in the cash register worth the price of having to live with taking the life of a child. Imagine trying to live your life with a horribly traumatic scenario like this, day and night.

Understand The Responsibility

A firearm should never be used as a bluff or as a means to run off an unruly customer. You should not carry it out into the parking lot with you when kids get in arguments or burn rubber in the lot. Most states have laws against the brandishing of a firearm. In some states, that law is called *assault with a deadly weapon*; it is a felony, and will get you state prison time.

Ask yourself—is there anything in the register worth dying for? Anything there worth killing someone over? One ends your life and the other changes your life—*forever*. Few cops ever are the same after having to take the life of another human being.

Storing Your Weapon

If you're going to keep a gun at your business, you may want to have it hidden in a back room. Again, if a robbery suspect is going to kill, they will frequently take the victim into a back room (as we discussed in the Armed Robbery chapter). So if you are being directed to go to a rear room, the situation has just become much more serious. Have a box with a few bucks in it hidden somewhere near the gun. Offer to get the money for him. If he allows you to get the money, you might have a chance to go for the gun. Try grabbing the box with one hand and the gun with the other (hoping he'll focus on the box in the one hand and not the gun in the other) and turn, so that the first thing he sees is the box. This will hopefully give you time to fire. Again, you shoot to stop, and that spot is *center mass*—the center of the chest. If the suspect tells you to back away and that he'll get the money from the box himself, make certain that he won't be able to see the gun from that location. Realizing that he'd almost been had, he may turn violent.

In Conclusion

If you have a gun or guns in your business, have them hidden and also have a plan for getting to them. Most importantly, know the laws in your state concerning the use of guns, become proficient with the operation of your weapon, and ensure that any employees who have access to or permission to use the gun are also trained. This is a highly litigious area, despite the right to defend one's self. Recognize the dangers inherent in taking the offensive as well—know that any risk you take may cause your own death.

Chapter Twelve

Robbery And Burglar Alarms

If you have a robbery alarm, use it only for robberies. It is not a tool to report traffic collisions that occur at the intersection; it should not be used for petty thefts or individuals trying to pass bad checks or minors trying to purchase beer. It is only to be used in the event of a robbery. When officers are dispatched to an armed robbery in progress, they respond without any undue delay—meaning, they *scream to the scene*. How would you feel if, while racing to the scene, either the officer or some citizen was injured as the result of a traffic collision. Certainly the officer must drive with care, and both he and the agency will accept responsibility for any failures in that department. But if you used your robbery alarm to report an argument in the parking lot, you will most likely find yourself involved in civil litigation. In some cases, you might even be criminally charged for the false report of an emergency.

Know How To Operate Your Alarm

Many agencies have instituted a system of fines for false alarms that can compound and grow to several hundred dollars for a single malfunction. In many locations, the police may be able to withdraw your permit to have such an alarm, and then you are left with nothing. Generally, it is not the business owner, but the newer employees who set off the alarm accidentally. Make the operation of the alarm part of your employee training program. Make sure that your training program includes an explanation, a time for the employee to observe you setting and deactivating the alarm, and then a period of time where the employee does the manipulation of the alarm, while being observed by you or a trained employee.

Determine from your alarm company whether or not your alarm goes to an answering service, which then calls the police, or whether it goes directly to the police. Additionally, find out if your alarm has its own dedicated line or if it is linked through your phone line. If it is linked through your phone line, once you have activated the alarm you will not be able to call the police because the line will be tied up!

How Police Respond To Alarms

When officers roll on a robbery alarm, they treat it as real (even though most aren't—it's a survival *must*). They come prepared for an armed confrontation and hope that it won't occur. Their guns will be drawn and anyone leaving the premises will be "prone'd out." This means they will be commanded to lay face down on the ground, asphalt, or wherever they happen to be when the officer arrives (they will not appreciate your improper use of the alarm, either, and just may make issue of it—through their attorney). The worst fear of any officer is to shoot someone unnecessarily. If you use your alarm to report a possible drunk subject trying to buy more beer, and upon arrival an officer orders him to stop and he refuses—a scenario is set for a possible shooting, where someone could unnecessarily die. You'll get to know the officers involved when you go to civil court with them. Then, you have to find a way to live with what you have caused.

Purchasing An Alarm System

Where burglary alarms are concerned, be sure you buy a good one. Demand quality service from the company that installs it. Not only might you be charged for responses to false alarms (and perhaps lose your permit to have an alarm), but additionally, you may become a very unpopular business owner. Many times, law enforcement agencies are holding a myriad of calls and are constantly prioritizing and re-prioritizing them. No officer enjoys handling your burglary alarm (which is probably false, just like the other fifty times it has activated), when there are matters pending that are more likely to be real crimes.

Even if you set off your own alarm and call right away, most agencies have to send an officer to your business, anyway. There's always the fear that they may be talking to the burglar and not you, in which case the agency will be civilly liable for your loss.

A strong business relationship between you and the alarm company installing and maintaining your system is of utmost importance. You must demand that the system be configured so that it functions effectively. A slight wind should not set it off, neither should turning on the air conditioning. These excuses do nothing to prevent one or two officers from having to respond to and check your entire business. Your alarm installer can utilize various types of detectors to prevent these erroneous activations and can adjust their sensitivities.

Be certain that the police agency or your alarm company (or both) have up-to-date name and phone numbers for call-out personnel. If you or your assigned employee refuse to respond when called on an alarm activation, the police most likely will still check the location, but will not be held accountable if you arrive in the morning to find out you've been cleaned out. There are many areas law enforcement cannot check (rooftops, interiors) if you fail to respond with a key. Refusing to respond sends a very clear signal to most law enforcement officers as to how much you care whether or not you are ripped off, even though this may not be the total truth. Respond and respond quickly! If it's going to take you thirty minutes to get there, do not tell them fifteen. Provide the dispatcher with a description of the vehicle in which you will be arriving. It helps officers at the scene identify you and removes any concern about who is arriving on the scene.

Also, make sure that your alarm company provides you with a legend of the alarm's panel information. Seeing that "Zone One" has activated means absolutely nothing if you cannot tell the officers where "Zone One" *is!* It is infuriating for an officer to roll on a burglar alarm and watch a business owner try to figure out whether the activation is a perimeter alarm, a safe alarm, a temperature alarm, or a low-battery indicator. Buy a good alarm from a reputable alarm company that provides quality service and ensures that you are trained in its use.

In Conclusion

Both burglary and robbery alarm types are discussed more thoroughly in the chapter entitled "Technology For Prevention/Detection." It is important enough to be redundant, however, as long as it is stressed that such alarms be used *only for their intended purpose.*

Chapter Thirteen

Technology For Prevention/Detection

Camera Systems

It's all there—anything you could possibly want. Where technology is concerned, *there's always a place to put your money.*

There are video board cameras that fit in fake speaker boxes, fake electrical outlets, and fake smoke detectors. There are video cameras with pinhole lenses that work on drop ceiling mounts—cameras that "detect" motion and follow it. There are recording decks that run A/C or D/C, have time-lapse capabilities, and read out the date and time on the video image. Some units allow you to obtain photographs directly from the video image via a built-in printer. Most of the quality systems allow the viewer to switch between cameras and pan and tilt the camera, as necessary, to keep a watch on things—and on people. All of this can be done remotely, from some cubbyhole in the back part of the store.

There are still image cameras that fit within fake speaker boxes, fake radios, and behind trick Plexiglas, known as "black glass" (an "it-can see-out-but-you-can't-see-in" arrangement). These still cameras may be activated by: hardwire systems attached to "bait money" and placed in

Technology For Prevention/Detection

a micro switch right inside the register; softwire systems attached to transmitters that fit inside a paper money slot in the register; and remote transmitters, carried like jewelry on a chain around the neck. When activated, their built-in, auto-winder allows the camera to take a series of as many as fourteen photographs—kind of a play-by-play recording of a crime.

In addition to elaborate hardwire alarm systems, one can purchase portable units that will allow them to be moved from store to store and from franchise to franchise. These alarm systems can be activated by a trip wire, a positive- or negative-pressure pad, or by a transmitter. They can activate video and/or still camera systems and even transmit a taped message (that you record) to your alarm company or home. They can be interfaced with an auto-dialer and upon activation, automatically linked to your home, alarm company, security service, or if your law enforcement agency will allow, their switchboard or alarm board.

Other Anti-Theft Devices

Once again, if you can think of an application, there is a company somewhere that makes the product. In this case, its more a matter of how obvious you wish to be, rather than how much money you have to spend.

Let's consider the age-old problem of an employee stealing from the boss's desk. Whether it be money, products, or information, the problem is still the same. Announcing the investigation will insure failure—the employee will simply stop stealing until things cool down, or change the method used. On the other hand, no business owner wants to confront all employees and interrogate them. The truth is, this would most likely be of little value anyway. All too often, the one who's ripping you off is the one you'd least suspect.

It's important that preliminary studies be made prior to the initiation of any of the techniques that will be discussed. The purpose is to narrow down the field of possible suspects. Are the thefts occurring on a particular day? On a particular shift? During these "frames of opportunity," who has access to the object (or objects) of the thefts?

The products available must be divided into two classes: *active* and *passive*. The one used depends upon whether or not you already have an idea of who the culprit is.

Technology For Prevention/Detection

Passive class products require that, after their employment, the fact that they were used be disclosed. For example, one product places a residue on the surface contacted (desk drawer, cash box, etc.) that remains on the hands of anyone who comes into contact with the item. It cannot be seen (and therefore no effort is made by the culprit to remove it) except under ultraviolet illumination. The material comes in the form of a powder, a liquid that dries, and even a marking crayon. Obviously, a small ultraviolet lamp source must be purchased. The problem is, that you have to check everyone for the presence of the residue, and your clandestine operation is now well-known.

Active class products produce an immediate reaction. One such product, known as "Anti-Thief Stain Detection," comes with four or five different colored powders (you match the powder color to the surface you intend to place it upon) and a liquid that dries into a clear state. When contacted by the suspect, the material stains the hands a bright color (usually pink or purple). Since it reacts with moisture (perspiration), attempts to wash it off will generally cause it to become more visible. A persistent crook can, however, scrub the majority of it off with scouring cleansers, but generally fails to get it out of the cuticle areas and flexures of the fingers. More often than not, the employee who discovers her hands are stained will not even know where she contacted it. She may come to work as usual and laugh about it. Or if she suspects she's "been had," she may call in sick. Obviously, this would be an excellent time to drop by and see how she's doing. One drawback of the staining powders is that it is difficult to clean up. You might consider placing it on a single document or dollar bill—on something which you could toss out without great loss.

Is someone stealing fuel? They make highly-concentrated liquid stains that will stain thousands of gallons of fuel. Then, all you have to do is siphon a little gas from the suspected employee's car.

Reasonable Solutions

Expensive video surveillance systems are generally the choice of major department store chains. For the small business owner, it is necessary to select from a variety of lower-cost (but not necessarily lower-value) tools.

Where technology is concerned, the only limit is your budget. One can spend anywhere from twenty-five dollars for a convex observation mirror to thousands of dollars for complex video surveillance systems. The key is to spend the least amount of money for the maximum effect.

Technology For Prevention/Detection

Crime prevention, technically, cannot be measured. Although you can compare the number of shoplifters caught one year against the next, you can never know how many crimes your actions prevented. Where thefts are concerned, psychological warfare is an inexpensive alternative.

A sign placed in the window of your store, adjacent to the door, that says: "We prosecute shoplifters to the full extent of the law," sends a clear signal that you are theft conscious—in other words, they know you'll be watching. And if you're concerned about the effect it might have upon your law-abiding, good customers, consider adding the following: "In order to provide the best possible prices for our customers, we prosecute shoplifters to the fullest extent of the law." Such a sign represents a minimal investment for the store owner.

Convex mirrors are a wonderful tool for allowing you to see into otherwise hidden areas of the store. They also will cause some shoplifters to go to another store. Again, they show the would-be thief that you are theft-conscious.

Most electronics stores offer reasonably-priced, electric eye light beams that operate in conjunction with a bell or chime. When placed at the door, you are "notified" by sound every time someone enters the store. This gives you the opportunity to come from behind the counter (or from the back room) to make contact with the customer. One of the best ways to prevent theft is to look the potential thief in the eye and say, "May I help you?" They'll leave almost every time—without stealing from you.

For under fifty dollars, you can purchase dummy video cameras that come with a ceiling mount. Many actually plug into a 110-volt outlet and have a red L.E.D. that is easily seen by anyone looking. They appear very real.

Alarms and Alarm Activators

It is always advantageous to have a robbery/burglary alarm at your place of business. You do have to decide, however, just what you would like this alarm to do for you.

With burglary alarms, you may opt for a local bell, which rings loudly at the location. You will most likely limit your losses to a damaged door or a smashed window, and a few items that can be quickly carried from the store to a waiting vehicle. The chances of the police apprehending someone, however, is not likely—that loud bell will send them

Technology For Prevention/Detection

flying from the scene. Another option is a silent alarm, which sends a signal to a security service or directly to your local law enforcement agency. The hope is that law enforcement will be able to arrive at the scene and catch the perpetrators either in or exiting the store. Thus, you not only recover property, but also take two burglars off of the street. If there is a delay in response (due to slow notification by the alarm company or because the police units were extremely busy or out of position to respond quickly enough), you may suffer a greater loss. A third option is to have both—an alarm that sends a signal to the alarm company or law enforcement agency *and* rings a loud bell at the store. It is hoped that this set-up will minimize losses and, if the cops happen to be close, maybe catch a crook or two.

If you cannot afford an alarm system, do not assume that you can't still have the benefit of one. A bell box (the metal box with the louvered door) placed under the eaves of the business and carrying an alarm company sticker on it may prevent you from being burglarized. Many burglars have explained to detectives that they purposely avoided a particular residence or business, simply because they saw a bell box. Why break into an alarmed location when you can go to another one that has no alarm?

If you have an alarm system, whether it be linked to an alarm service or the police agency, be certain that the names and phone numbers of call-out personnel be updated continually. Again, there's nothing worse than having a business that has been burglarized and cannot be secured, simply because police cannot locate an owner or employee of the business.

Alarms can be activated through a variety of means. Interior motion detectors are excellent in cases where someone has hidden in the business prior to closing. Since they were already in (and didn't have to break in), the perimeter alarm would not be activated. In fact, if all you had was a perimeter alarm, the only activation that would occur would be when the perpetrator left the business. With motion detectors, the "stowaway" is quickly detected the minute he moves. These also work quite well to prevent roof entries when they are mounted in the attic. The downside of motion detectors is that they can be activated by moving displays, aisle identification signs that are caused to move by the air conditioning, and even by some cat that sneaked its way in.

Heat sensors also work well. These devices are able to detect body heat. The sensor is set to activate and send the signal at a particular temperature—98.6 degrees is generally warmer than most proprietors keep their businesses, even when closed.

Technology For Prevention/Detection

Trip wires are sometimes used. They require that the suspect traverse whatever "alley" across which you have stretched the activation line (usually monofilament fishing line). In a large store, it is possible that the suspect might never pass through this location. If set up correctly, the line is generally not felt when coming in contact with the lower legs, and thus, the suspect seldom knows he has activated the alarm.

Activation pads are quite effective. In this case, a rubber pad is placed onto the floor (generally near the register). When the suspect steps onto the pad, the pressure activates the alarm. When the store is open and the alarm is off, the pressure pad appears to be a rubber mat, placed by the register for the clerk's comfort.

There are also pressure pads that come as a strip. These are placed underneath the edge of the counter. When a robbery occurs, the clerk need only place pressure on the pad to activate the alarm.

There are also alarm transmitters available. Some fit on the belt like a pager, while others may be worn around the neck like a necklace. Again, in the event of robbery, the clerk need only depress the activation button and the alarm will be sent.

The standard burglar alarm system consists of electricity-conducting tape, placed upon the windows, and magnetic contacts which are placed on all doors (and in some cases, windows). If the suspect uses a glass cutter and is able to remove a sufficient amount of glass to allow entry, such an alarm may never activate. (A motion detector installed in the interior would foil this attempt, however). Owners of tape/magnetic contact alarm systems should insure that the minor movement of the door does not set off the alarm. Otherwise, you'll have police responding every time some patron arrives after closing, thinking that you're open, and tries to open the door. False alarms in many jurisdictions cost money.

There are also many devices on the market that are designed to assist in the identification of the suspect. Also, in some cases, these devices will render the product of their crime useless. For example, there are exploding dye-packets that are often used by banks. The dye-packet explodes shortly after the suspect leaves the bank, staining the perpetrator and the money. There are also hand-held devices that shoot a non-lethal projectile which explodes upon contact, sending dye-stain all over the suspect. One must be constantly aware, however, that actions of this type may cause an even more violent reaction from the suspect.

There are a wide variety of alarms and an even greater variety of activation devices. What you should purchase and employ is dictated by your specific needs.

Chapter Fourteen

Bomb Threats

Introduction

Businesses periodically receive bomb threats, which place business owners in a difficult situation where they must balance the possibility that the threat is valid against the statistical data that would suggest otherwise.

Bomb threats may be made to stores for a variety of reasons. With the proliferation of hate crimes in America, we are seeing an increase in bomb threats directed toward businesses owned by new immigrants to this country. Places of worship are not exempt, either. Threats may be made by disgruntled employees who were denied pay increases or promotions and because of this, seek to disrupt business and hopefully create a loss of revenue for the owner. Included within this category are persons who were fired; the reason for the firing isn't really a major factor. What is important is what the fired employee perceives. If she feels the firing was unjustified, even though knowing she was guilty of the offense(s) identified, or if she feels humiliated by being fired in front of other employees or customers, she may resort to a bomb threat for revenge.

Bomb Threats

What To Do If You Receive A Bomb Threat

If you should receive a bomb threat over the phone:

1. Note the time when the call was received.
2. Note the time when the caller hung up.
3. Try to remember the exact words of the person making the call.
4. Did the caller refer to anyone by name? Did the caller refer to any particular event? Condition? Situation? Belief?
5. Is the caller's voice male? Female? Disguised?
6. Does the caller's voice sound young? Middle-aged? Old?
7. Is the caller's tone of voice calm or excited?
8. Does the caller's voice sound practiced or is the caller making up what is being said as she goes along?
9. Is there a discernible accent?
10. Is the voice muffled, as if something has been placed over the phone?
11. Is there any background noise? (Traffic, music, baby crying, other people talking, etc.)
12. Does the voice sound familiar? Who does it sound like?
13. Were there any threats that were related to race, nationality, religion, or type of product sold? (Pornography, pipes used for smoking drugs, etc.)

Try to ask the caller:

1. When is the bomb going to explode?
2. Has the bomb already been placed? If so, where?
3. What kind of bomb is it?
4. What does it look like? (Suitcase, shoe box, etc.)
5. Why are you doing this?

Once the call is concluded, note the time and the line upon which the call came in. Is this number available to the general public? Or could only someone with insider knowledge have known this phone number? By dialing this number, was the caller assured of getting the specific person she wished to speak to, i.e., the manager? Would this number be available to the general public and advertised as the manager's line? Is there anything to indicate that the call may have originated from inside the store? (Sometimes interior lines have a different sound than those calls coming in from the outside.) Did the caller seem to have any information that only an employee would have known? Did the caller mispronounce or misuse certain words?

Bomb Threats

Immediately notify the police. Inform them of what the caller said, including any time lines provided, and describe the premises (multistory building, day-care center, etc.) and surroundings. If there is any reason that you believe the threat may be a valid one, be sure to pass this along, as well.

The decision to remove all customers and employees is a difficult one. Most threats are just that—mere threats. In this country, (thankfully) we have not had near the experience with actual bombings as in the Middle East, Northern Ireland, and other such locations where political ideologies, religion, or ethnicity have created a high level of psychoticism. In some jurisdictions, there are statutes that make it mandatory for police to clear the building and any exposed buildings of persons. The decision, in these cases, is made for you. In many others, however, there are no state or local laws covering what actions should be taken in such cases—the decision rests with you. If a disgruntled employee realizes that he can totally disrupt the business, he may continue to call in bomb threats. And yet if the threat is real, and no evacuation is initiated, you may be civilly and ethically liable for any injuries that are sustained. Police, in most cases, will not recommend a course of action—the decision belongs to the owner, manager, or person in charge.

Whatever you decide to do (keep the business open and search for any device or clear the premises), the police will assist you. In multi-story buildings, it is necessary to clear not only the entire floor the bomb is reported to be on, but also any floors above and below. Failure to notify persons, in other areas of the structure, of the threat will most likely be adjudged as negligence should the device actually have been planted and then detonate. Move people well away from the buildings in case there is a detonation. Flying shards of glass, metal window frames—*almost anything*—could be fatal.

Identifying Possible Bombs

Trying to find a bomb in a store is like trying to find the proverbial needle in the haystack. A bomb can look like most anything, be placed in most anything, and be activated by a variety of means. The entire store would have to be searched, primarily looking for something that appears out of place. Check bathrooms, dressing/locker rooms, and wastepaper baskets. If something suspicious is found, do not move it—leave it for the police bomb experts. Some agencies have dogs trained to sniff out bombs, but even then, depending upon the size and design of your business, this may not be a practical method.

Bomb Threats

Types Of Bombs

Pipe bombs are the most common device used. The necessary parts can be obtained at a hardware store and the black powder at any gun shop where reloading supplies are sold. Certainly, locating any section of pipe, be it metal or PVC, where both ends are capped and a wire is protruding, should be treated as a bomb discovery. However, all the bomber would have to do is place this device in a shoe box and put it on a shelf and it may go undetected for years (assuming it never detonated). Dynamite, TNT, and hand grenades are most assuredly available if one looks hard enough—anything can be purchased for the right price. C-4, a plastic explosive, is common to the Middle East, and Semtex, another plastic explosive, is manufactured in Czechoslovakia. Although not common to the United States (except in military operations), plastic explosives have been discovered here.

Bombs can be activated by a variety of means. Simply lighting a fuse is not a logical method due to the lack of escape time for the bomber. However, delayed fuses have been fashioned out of a cigarette and a matchbook or using stick incense. Pressure activation, both positive and negative, are another means of activation. With positive pressure, the bomb is detonated by the act of stepping upon the activation device, which sends a signal either through air pressure or electricity. A negative pressure device activates when something heavy, placed upon the device, is moved (such as a box, a vehicle rolling forward, etc.). Radio signals may also be the method of activation chosen by the bomber. That is why most intelligent police officers shut off their portable radios (handhelds) once they have arrived at the scene. It is not the receiving of radio signals that is necessarily the problem, as the air is constantly full of radio signals. It is when the transmitter is "keyed up" at the location that the greatest risk occurs. And of course, there is always the chance that the bomber has a sophisticated transmitter that sends a signal to the receiver attached to the bomb to create a detonation (although this is quite uncommon and much too sophisticated for most).

Bombs are designed basically to produce one of three effects: (1) fragmentation; (2) blast pressure; or, (3) fire. Fragmentation is probably the most common method. The rapid expansion of the gases inside the tightly closed container (such as a pipe bomb) causes the container to rupture, sending out shards of metal at an incredible velocity. Even if the bomb container itself would not create major fragmentation, the objects placed inside may. Bombers have been known to fill the space with nails, screws, ball bearings, and the like. In some cases, simply the object that the bomb has been placed within may be the source of the highly injurious fragmentation (metal desk, display, etc.). With blast pressure devices, the concussion created by the explosion is enough to break bones. The pressure created turns everything into fragmentation. The incendiary device is constructed for one purpose: to create fire and burn and maim anyone close enough to the device or caught within the structure as it burns to the ground.

In Conclusion

The bottom line is that you cannot easily identify a device, nor can you totally protect yourself from harm when searching. Keep this in mind when making any searches along with police officers. Always identify items that are out of place and don't belong, but don't touch them. In one case, a store owner told a police officer that he need not be concerned about the wine bottle box found on the office floor as he had "already picked it up and shook it."

It is also possible that managers, department heads, or business owners may receive a bomb via the U.S. Mail. Be wary of any mail that does not have a return address on the envelope. Be wary of any mail that appears to have any staining or the presence of any oily residue. The explosive material is activated by the mere opening of the envelope.

Index

A

abduction 29
absconding 12
abuse 34, 57
abuse of sick time 34
accomplice 24
account balances 28
accountant 14, 39
accounting 41
accounting procedures 27
accounts 13, 27, 34, 38, 41, 77-84
acquittal 58
addict 22, 39, 73
addiction 44
advertising 67
airport police 3
alarm boxes 50
alarm button 62, 63
alarm permit 68
alarm signal 62
alarm systems 75, 113-115, 117-122
alcohol 16
alcoholic beverage enforcement 7
amount in controversy 10
amphetamines 22
anonymity 37, 50
answering service 62
Anti-Thief Stain Detection 119
appeal 9
appeals courts 3
appeals process 9
appearance 29, 44, 46
appellate courts 9
applicant's appearance 17
applicant's home 20
applicant's lifestyle 17
applicant's work history 17, 19
applicants 17-21
apprehension 43
armed robbery 28, 29, 61-72
Arrive, Leave, and Return 45
arson investigators 3
assault with a deadly weapon 110
asset protection 14, 43
assets 17
assistant state's attorney 7
attorneys 7
audit 27, 28, 41
auditors 14, 27
authority to make the arrest 53, 54
authority to purchase 33
authorization to perform 2
auto theft 6

B

background checks 20
background investigation 17, 21
bad credit 17

Index

bail 10, 11
bail bonds 11
bail bondsman 11
bank deposits 71
bankruptcy 16
becoming a hostage 68, 69
behavior 21
believability 59
best evidence rule 80
blank checks 28
bomb threats 37, 123-127
bondsman 11
bonuses 20
bookkeeper 23
booster box 46
bounty hunters 11
Bureau of Alcohol, Tobacco, and Firearms 2
Bureau of Engraving and Printing 85
burglars 32
burglary 6, 16, 25, 73-75

C

cameras 25, 50, 51
case number 58
cash drawer totals 26
cash on hand 29
cash payment 26
cash refund 40
cash registers 25, 27, 32, 62, 65, 67, 74
cashier location 49
cashier's station 49
cattle theft 6
chambers of commerce 10
charge accounts 93-97
Check-Alert Services 84
check forgery 79
check out system 23, 24
checks 27, 28, 77-84
checks and balances 28, 77-84
citizen's arrest 52-54, 56, 58
city attorney 7
civil 6
civil actions 10, 78
civil courts 3, 9, 10

civil disturbance 4
civil lawsuit 52, 56, 58
civil liability 44
civil litigation 33
civil papers 10
civil suits 35
civil trial 10
clerical error 27, 29
cocaine 16, 22
cocaine paraphernalia 22
communications 5
communications division 4
community relations 4, 107
complaint takers 4, 5
complaints 35
computerized fingerprint system 84
concealed store property 51, 55
concealed weapon 54
concealing property 45, 46, 50, 58
concealment 51
conclusive evidence 29
confession 33, 36
conflict of interest 9
constable 10
constant nervousness 35
containers 55
convenience markets 29
correctional institutions 10
corrections 3, 10
corroborating observations 55
counseling 11
counterfeit cash 2, 85-90
counterfeit products 91, 92
county 2, 3, 10
county jails 10
county sheriff 3
court 8, 9
court of last resort 9
credibility 57, 59
credit 17, 94-99
credit application 95, 96
credit bureau 17, 95
credit check 95
credit history 17

creditors 96
credit risk 95
crime lab 6
crime prevention 4
crime scene 6
crimes against persons 4, 5, 6
crimes against property 5, 6
criminal complaints 35
criminal courts 3, 9
criminal histories 58
criminal intent 77, 78
criminal investigation 6
criminal justice system 1, 9
customers 14, 22, 24, 26, 33, 38

D

D.A.R.E. 4
daily receipts 24
darkness 61
deadly force 69, 70
declaration of arrest 52
declining medical treatment 31
dedicated line 62
defense attorneys 7, 8, 59, 60
defense counsel 7
delay tactics 56
deliveries 26, 29
delivery drivers 25, 26, 31
demeanor 45
Department of the Treasury 2
dependency 21
deposits 71
deputy district attorney 7
description of suspect 30, 31
detainment 51
detective 6
detective division 6
detoxification centers 10
dilated pupils 22
direction of travel 62, 64
discount 39
disgruntled employees 34, 37
dishonest employees 14, 34, 39
dispatcher 4, 5

Index

display aisles 49
display items 49
displays 49, 67
district attorney 7
diversionary tactics 69
division of responsibility 3
double standard 40
drive-up window 71
drop money 62
drop policy 62
drug abuse 21
drug addict 73
Drug Abuse Resistance Education Program 4
drug habit 33
drug screen 22
drugs 12, 21, 22
due process 9
dumpster 25, 32, 41

E

electronic safes 62
embezzlement 13-41
embezzlement of cash 25
employee receipts 26
employee theft 16, 23
employee-consumer contact 49
employees 14-21
employment screen 21
energy swings 21
equipment for personal use 17, 23
ethics 16
ethnicity 31
evictions 10
evidence 5, 7, 25, 29, 58
evidence suppression hearings 9
ex-employees 37
expense accounts 26
explanations 57, 59
exploding dye-packets 122

F

failure to mark items 26
false identification 81
false pregnancy 47
false report 28, 33
false robbery 28, 29
falsification of evidence 58
family court 3, 10
Federal Bureau of Investigation 2, 14
federal courts 9
federal facilities 10
federal marshals 10
federal prosecutor 7
females arrested 53, 55
fictitious employee 27
fictitious name 27
field officer 5, 6
field patrol 5
fingerprints 65, 66, 83, 84
firearms 69, 70, 109-111
fixtures 32
flash-bang/concussion grenades 69
focus on personnel 45
follow-up 6
follow-up detectives 6
follow-up investigation 7
For Deposit Only 32
force to effect the arrest 54
forensic examination 6
forgery 27
forgery/fraud 6
formal confession 33
fragmentation 126, 127
fragmentation bombs 126
frame of reference 65
fraud 28, 39
friends 20, 24, 27, 35
functional appearance 51

G

gambling 16
generic bills 23
gifts 23, 33
goals and expectations 41
government agencies 19
grab and run 74
guilt 7, 33, 36
guilty pleas 44, 58
gunfire 68

H

habits 18, 21, 73
half-sphere camera mounts 51
handwriting 40
handwritten receipts 26
harassing witnesses 8
harassment 8
harbor police 3
heat sensors 121
Hepatitis B 31
highway patrol 3
hiring friends and relatives 20
HIV 31
holidays 29
homicide 6
hostage situations 68, 69
housing authority police 3
hypersensitivity 30

I

identification 8, 79-81, 83
identification cards 81
identifying characteristics 65
illicit drugs 16
inconsistency 30, 33, 35
inconsistent statement 8
incorporated area 2
initial report 2
initials 48, 58
injury 22, 29, 34
inkless fingerprint system 83
inside information 19
internal audit 28
internal theft 17, 28
interviews 17, 35
intimidating 8
intimidation 59
inventory 13, 14, 16, 23, 24, 34, 35
inventory systems 16
investigating agency 33

Index

investigation 5-7, 28, 30, 31, 33, 36, 37
investigators 3, 6, 8

J

job application 18
job injuries 34
judges 8, 9, 59, 60
judiciary 7
jurisdiction 1-3, 7, 10, 48, 52, 58
jurors 14
jury 8, 54, 59
justice 7
justice courts 9
juvenile court 3

K

keys 41

L

lab tests 22
labor complaints 35
lack of responsibility 17
landscaping 74
larceny 43
lawsuit 10
Levi-Strauss Company 92
legitimate victims 30
license plate 66
lighting 67
line-ups 32, 33
lockers 32, 33
Los Angeles Police Department 4
loss prevention 43
lower level courts 9
lunch boxes 41

M

magistrate's courts 9
Making The Arrest 52, 53
male shoplifters 55
manipulation of accounts 27

mark items 26, 40
Marshal's Office 10
mask 29
maternity clothing 46, 47
medical treatment 31
minor offenses 9
mirrors 45, 50, 51
miscellaneous crimes 6
misuse of the alarm 68
morale 41
morality 16
motive behind shoplifting 44
motor vehicles department 3
mug books 33
municipal courts 9
mutual friends 20

N

narcotics 6, 15, 22, 35
needle marks 22
negative advertising 14
negative amount 26
negligence 3
nervousness 30
night deposits 71
non-essential police services 5
non-working products 23
notes 36, 47, 56, 57
not-sufficient-funds 78, 80

O

OR'd 11
oath 58, 59
obligation to report 37
observations 36
obstructions 67
offensive weapons 54
ombudsmen 107
one-way mirrors 51
original jurisdiction 9
out-of-state checks 82
over-rings 40
overcharges 44
overtime 15, 34
overtime opportunities 34

own recognizance 11

P

packaging 45
palming 26
panning video cameras 51
paper trails 28
paradigm shift 16, 22, 39
paranoia 21, 22
parish 2, 3
parish police 3
parole 10-12
parole agent 11, 12
parole revocation 12
parolees 12
pat-down-type searches 54
patrol division 6
patterns 24, 27, 36
payroll clerks 27
peace officer 3
perceived favoritism 31
perception of time 65
perception of getting caught 15
perceptual skills 29
perjury 58
perpetrator 28-30, 32, 53, 57, 60
personal purchases 23
personal vehicles 32
personnel management 34
petty larceny 43
petty theft 43
photographic line-ups 6
pilferage 23
plaintiff 10
point of sale 26
Police Athletic Leagues 4
police jargon 57
police negotiator 69
police report 35, 39
police station 71
policy 24, 39, 40
poor service 44
Porelon® ink pads 83
position of trust 13
positive identification 67

Index

Post Office Boxes 41
potential targets 61
pre-trial publicity 3
precinct 2
preliminary hearings 9
preliminary investigator 5
preparations for stealing 45
presence of counsel 31
prison 1, 10, 14
private attorney 7, 8
private investigators 8
pro-active 5
probate courts 9
probation 3, 10, 11
probation and parole 10, 11
probation officers 8, 11
professional demeanor 59
professional shoplifters 44
profits 14
promise to appear 10
property taken 59
pro se 7
prosecuting attorney 59
prosecution 20, 33, 79
prosecutor 6, 8, 9
public defenders 7
public record 20
punishment 40
purses 40

Q

questioning 53, 59, 60

R

radio dispatchers 4
radio receivers 67
rear storerooms 32
receipts 26, 27
receiving reports 40
record-keeping 40
records 13, 28
references 19
refund schemes 48
rehearsed description 30
reimbursement 15, 27
relatives 20

remotely activated alarms 51
reporting thefts 37
reports 47, 56-59
responding units 63
response time 63
restaurants 26
restitution 40
returned goods 24
robbery 6, 28, 29, 31, 50, 51
robbery alarm 62, 63, 67, 68
robbery prevention 51, 67
robbery suspects 32
rolled coins 32
route 71
ruse 66

S

salary 19
sale posters 67
scams 24, 26, 40
scanning radio receivers 67
security devices 37, 50, 113-115
security mirrors 45
security officers 45
security personnel 25
self-injury 29
self-sequencing cameras 51
sequentially-numbered-receipts 26
serial numbers 75
serious offenses 9
sexual assault 6, 29, 57
sexual battery 55
sheriff's department 2, 3, 6
shipping and receiving 25
shoplifters 14, 40, 43-60
shoplifting 24, 43-60
shoplifting prevention 51, 60
shopping partner 67
short register 39
short ring up 23-25
shrinkage 23
sick time 34
signs of Cocaine Abuse 22
silent alarm 67
small claims actions 79, 80

snitch 37
social activities 35
Social Security cards 81
societal wrongs 3
specific action 57
state police 3
statement 6, 8
Stockholm Syndrome 69
stolen identification 80
store event number 58
subpoena 59
substance abuse 16
substance abusers 12
sudden prosperity 34, 35
suicidal subject 69
summons 59
supervision styles 16
Supreme Court of the United States 9
surveillance cameras 44, 119
suspect 6, 8
suspect description 62
suspicion 16, 36
SWAT 68, 69

T

tag-activated magnetic alarms 50
tail 71
tape 25, 26
tattoos 31
teachers 18
telephone 17
tension 35
termination 19, 33
territorial jurisdiction 2
testifying 58, 59
testimony 7, 56, 59, 60
The Means To Steal 46
The Stockholm Syndrome 69
theft 13, 15, 36, 37, 43-49, 51-55, 57, 59, 60
threats 37, 62, 63
till tap 48
time-lapse video recorder 25, 117
traffic courts 9

Index

transit police 3
trash bin 25
trash cans 25
trash disposal bins 40
trash enclosure 32
trauma 29
trauma level 29
travel 16, 26, 27
trench coats 46
trial 2, 3, 8-11
trial courts 9
trusted status 16

U

U.S. Small Business Administration, 6
unauthorized disbursements 28
uncompensated overtime 15
under the influence 22, 53
unexpected audits 41
uniformed patrol 5
unincorporated area 2
United States Postal Service 2
unknown assailant 31

V

vacation 34
vague terminology 57
VCRs 49
vehicle description 53, 57
verbal contract 78
verification 18, 26
video cameras 25, 50, 51, 67 117
video surveillance systems 50, 51, 119

W

waiver 17
weapons 29, 54, 61-63, 67, 70, 109-111
wholesale 23
wide angle mirrors 50
winging 70

witness stand 59
witnesses 4, 8, 9, 30
work history 17, 19
working alone 29
written law 3
wrongful termination 33

Z

zero-tolerance policy 39

Grow Your Business With These Small Business Books From Sourcebooks—

Smart Hiring For Your Business by Robert W. Wendover
Everything you need to know to find and hire the best employees.
200 pages, ISBN 0-942061-57-8 (hardcover) • ISBN 0-942061-56-X (paperback)

The Small Business Start-Up Guide by Hal Root and Steve Koenig
Make or break factors to successfully launch your own business.
152 pages, ISBN 0-942061-70-5 (hardcover) • ISBN 0-942061-67-5 (paperback)

Getting Paid In Full by W. Kelsea Wilber   (March 1994)
Collect the money you are owed and develop a successful credit policy.
136 pages, ISBN 0-942061-71-3 (hardcover) • ISBN 0-942061-68-3 (paperback)

How To Sharpen Your Competitive Edge by Don Reynolds, Jr.
Finding, developing and capitalizing on crackerjack positioning!
200 pages, ISBN 0-942061-73-X (hardcover) • ISBN 0-942061-72-1 (paperback)

Protect Your Business by Sgt. James Nelson and Ofc. Terry Davis   (March 1994)
Top cops help you safeguard your business against shoplifting, employee theft, and more.
144 pages, ISBN 0-942061-69-1 (hardcover) • ISBN 0-942061-66-7 (paperback)

How to Market Your Business by Ian B. Rosengarten, MS, MPH
An introduction to the tools and tactics for marketing your business.
152 pages, ISBN 0-942061-48-9 (hardcover) • ISBN 0-942061-45-4 (paperback)

How to Get a Loan or Line of Credit for Your Business by Bryan E. Milling
A banker show you exactly what you need to do to get a loan.
152 pages, ISBN 0-942061-46-2 (hardcover) • ISBN 0-942061-43-8 (paperback)

Your First Business Plan by Joseph Covello and Brian Hazelgren
Learn the critical steps to writing a business plan.
152 pages, ISBN 0-942061-47-0 (hardcover) • ISBN 0-942061-44-6 (paperback)

The Small Business Survival Guide by Robert E. Fleury
Learn to manage your cash, profits and taxes. Plus, **No Entry Accounting**—an easy way of doing and understanding your own accounting without double entry bookkeeping.
256 pages, ISBN 0-942061-11-X (hardcover) $29.95 • ISBN 0-942061-12-8 (paperback) $17.95

Small Claims Court Without A Lawyer by W. Kelsea Wilber, Attorney-at-Law
File a claim and get a judgment quickly and economically. Written in clear, uncomplicated language, it includes details about each state's small claims court system.
224 pages, ISBN 0-942061-32-2 (paperback) $18.95

To order these books or any other of our many publications, please contact your local bookseller or call Sourcebooks at 1-800-798-2475. Get a copy of our catalog by writing or faxing:

Sourcebooks Trade
A Div. of Sourcebooks, Inc.
P. O. Box 372
Naperville, IL 60566
(708) 961-3900
FAX: (708) 961-2168

Thank you for your interest!

Would You Like to Be In Our Next Book?

We're compiling a series of books that feature the best small business ideas. These can be ideas to save money, make money, motivate employees, increase sales, market your service, take care of your equipment needs, finance your business, collect money from creditors, negotiate a lease, improve customer service, develop team management, recruit new employees, develop new products, manage your sales force...in short, *ideas that really made a difference to the way you successfully run your business.* **We'd love to include your ideas!** We will, of course, give you credit for any ideas we publish. *Submissions become the property of the publisher.*

Please send your suggestions on your letterhead to:

> **Attn: "New Ideas" Dept.**
> **Small Business Sourcebooks Series**
> P.O. Box 372
> Naperville, IL 60566

Please be sure to include your name, address, phone number, FAX number if you have one and any pertinent information about your business (for example, a description of your business, the number of employees, your position). And again,
Thank You for working with Sourcebooks.

A Note from the Publisher

Hello!

I'm Dominique Raccah, head of Sourcebooks, Inc. In our business book series we've tried to bring you easy-to-use material to help you improve your key business skills. We've developed this series, with the help of a lot of entrepreneurs like yourself, to tackle some of the more common problems in building a business. As updated editions of these books are published, your input will be invaluable.

Please take the time to jot down any ideas you may have about the books you are using:

- Was the book useful?
- What would you like more of? What would you like less of?
- What else should be in this series?
- What are the most difficult areas facing you as you grow your business?

Please make sure you tell us which book you were using at the time. And please include your name, address and phone number so we can let you know if we incorporate your suggestions.

Thank You For Your Time and Your Support.

With Best Regards,

Dominique Raccah
and the staff of

Sourcebooks, Inc.
Small Business Sourcebooks
P.O. Box 372
Naperville, IL 60566
(708) 961-3900   FAX: (708) 961-2168

Do You Service Small Business Clients Too?

We, at Sourcebooks, have developed our business book program to serve the needs of the small business community. We've gotten incredible response to this series. If you are an accountant, lawyer, banker or consultant who also serves the small business community and you think our books could be of use to your clients, please give us a call. We offer special discounts to people who purchase these books to give or sell to their clients. Please contact us at 708-961-3900 or write to us at the address listed above for details. We'd love to work with you.